Beyond Technology: Unlocking the Potential of Clinical Decision Support Through Adoption and Design

Jamie Olsen

Copyright © [2024]

Author: Jamie Olsen

Title: Beyond Technology: Unlocking the Potential of Clinical Decision

Support Through Adoption and Design

This book is a product of [Publisher's Jamie Olsen]

ISBN:

Contents

Chapter 1

Introduction

Health is a universal human aspiration and a lifelong concern of everyone. Health Information Systems (HIS) — Information Systems (IS) that collect, store, manage and transmit healthcare data — can play a vital role in health outcomes and health system efficiency (Fichman et al. 2011). However, HIS have thus far failed to transform health, the same way other IS have changed many different sectors, such as banking and air travel. Having practiced clinical medicine in low resource areas, I have always felt the need for information system support for the increasing cognitive load associated with delivering the right care to patients, whose number is increasing worldwide due to a variety of reasons. The ageing population and increasing prevalence of chronic conditions such as diabetes are factors responsible for increasing patient numbers (Atella et al. 2019). Existing IS theories and models, conceived mostly in the context of firms, cannot wholly explain the suboptimal adoption of HIS. HIS failure may have user and system factors (Heeks 2006), and being myself an e-health hybrid who transitioned from healthcare to information technology (IT), I wanted to explore both healthcare and IT aspects in the context of clinical decision support systems (CDSS). This study is an attempt to find why and how *"to navigate these dark waters [of CDSS design],"* as summarized by one of my study participants.

1.1 Background

The medical domain is complex and diverse. The nature of clinical decision making is involved, clinical knowledge creation is demanding, and delivering decision support at the point of care is challenging (Osheroff et al. 2007). HIS collect, store, manage and transmit healthcare data. CDSS are a particular type of HIS that "provide clinicians with computer-generated clinical knowledge, intelligently filtered to enhance patient care" (Osheroff et al. 2007). HIS, particularly CDSS, have a high failure rate (Anderson 2002; Olakotan et al. 2020) though there have been instances of success in specific domains (Chaudhry et al. 2006). It is widely believed that this phenomenon remains mostly unexplained (Belard et al. 2017; Castaneda et al. 2015; Kawamoto et al. 2005; Krawiec et al. 2020; Olakotan et al. 2020). I believe that an exploration of CDSS high failure rates will make a considerable impact on healthcare.

Clinical research brings forth frequent changes in evidence (actionable knowledge generated from a systematic review) and disease management guidelines. It is a challenge for young medical residents to keep track of ever-changing evidence and bring it to the bedside. From a Knowledge Management (KM) perspective, seasoned practitioners' tacit knowledge is challenging to externalize and codify. The evidence from the published literature does not reflect the personal experience of seasoned practitioners. Hence, evidence from clinical research captured in the biomedical literature fails to make a considerable impact on patient management (Reilly and Evans 2006). CDSS are vital for the application of clinical evidence to improve patient outcomes (Chaudhry et al. 2006) when clinicians grapple with the problem of information overload and suboptimal treatment outcomes (McGlynn et al. 2003). CDSS can potentially reduce medical errors (Bates et al. 2001) and improve the quality and efficiency of healthcare delivery (Teich and Wrinn 2000).

Most CDSS research projects focus on either the design or evaluation of artifacts. They do not complete the full circle of design science research that includes building, evaluating and communicating design artifacts (Rittgen 2009). The clinical domain, its complexities, and its impact are ignored, leading to statistically credible artifacts that are "suitable" yet "inappropriate" for clinical workflows (Riemer and Johnston 2014). CDSS need a reconceptualization of technology that considers clinician perspectives (Orlikowski 1992). An uptake of CDSS by the clinical community can potentially save lives.

Next, following Creswell's recommendations (Creswell and Creswell 2017), I briefly describe the research problem and purpose statement.

1.1.1 Research problem

Most CDSS fail despite providing sensitive and specific information. Failures may be due to the lack of consideration of clinical workflow and the context of use. This problem has been extensively studied (Dwivedi et al. 2011; Greenes et al. 2018; Wade et al. 2006), and various prescriptive recommendations have been proposed (Sittig et al. 2008). However, a comprehensive theory that takes into account artifact design and its adoption by the clinical community is lacking. To address this problem, I conducted a qualitative research study using the constructivist Grounded Theory (GT) method (Charmaz 2006), triangulating many data sources based on a novel clinical decision support system that I designed. I investigated this problem from both user adoption and system design perspectives by asking the overarching question: How do doctors describe and characterize their use of CDSS in their practices and their views on how CDSS should be designed?

1.1.2 Purpose statement

This qualitative study addresses the problem of low adoption rates of CDSS among clinicians. As mentioned above, I used a constructivist GT method (Charmaz 2006), which is a type of research design in which an understanding of the phenomena emerges from the data collected and the reality co-created by the researcher and participants. The constructivist approach to investigate CDSS use is rare, and the efficacy of CDSS is typically assessed using randomized control trials (RCT). Qualitative data were obtained using semi-structured interviews, along with the collection of clickstream data and participant demographics. The Theory of Planned Behaviour (TPB) was the sensitizing theory that influenced the interview guide and system design. The triangulation of rich data provided insights into the knowledge needs of CDSS users. Analysis of the data also provided prescriptive design guidelines.

1.1.3 Theoretical sensitivity

In a constructivist GT, the researcher's theoretical sensitivity is vital in co-constructing reality with participants (Glaser 1978). I have practiced medicine in various regions that are economically, culturally, and geographically diverse. I switched from healthcare to IT and electronic health (e-health) and designed HIS and CDSS predominantly for resource-poor regions. Because of my multi-disciplinary training, I understand the technical and clinical challenges in the user adoption of CDSS. I have previously evaluated other CDSS systems clinically to assess its accuracy (Eapen 2005; Eapen et al. 2007; Sáenz et al. 2018). From my previous studies, I have realized that accuracy is not the only factor responsible for the user adoption of CDSS. Hence, I adopted a constructivist approach in this study to understand the challenges faced in the real world when using CDSS.

1.2   Delimitations of scope and key assumptions

CDSS can support patients in self-decision making, such as whether a mole needs further evaluation by a doctor. This study focuses on systems supporting doctors' decisions, owing to my interest and experience with such systems. I explore CDSS in the complex context of healthcare delivery, and as such, some of the terms and concepts that I use may need clarification. Some of these are used by the layperson and have different meanings in different contexts.

Health is the state of being free from illness and injury, and disease is an altered state of health (Saracci 1997). In the disease state, the patients – the persons with the disease state   display a set of characteristics that they themselves describe (symptoms) or those trained in the domain elicit (signs). Diseases are treated (or managed) by 'interventions' that include medicines (agents that prove a benefit in a randomized control trial) or surgery. Disease states are labelled for easy description, identification and management, by representative names based on the symptoms, signs, interventions and a variety of other associated factors. A diagnosis is such a disease label ascribed to a patient based on his/her signs and symptoms (and other associated factors), and this is usually done by domain experts (Calvo et al. 2003). I use the term diagnosis also to represent the process of making a diagnosis. In this context, the correlations of signs, symptoms and interventions with these disease labels are called 'evidence' and are usually obtained by the 'systematic review' of published articles on the disease.

I use the term 'doctor' to represent the trained domain experts who make a diagnosis, though, in practice, some who do not use the title of doctor also make diagnoses (nurses, nurse practitioners etc.) The management of the patient depends on the diagnosis. In many cases, an exact diagnosis is not possible (and not necessary for management), and the doctor makes a list of probable diagnoses called a differential diagnosis (DD) (Cook and Décary 2020). The diagnosis is made based on evidence. However, in some cases, doctors identify correlations based on their experience that may not be fully codified or documented. I call such undocumented, noncodified tacit awareness of doctors as 'knowledge' to distinguish it from evidence. The signs and symptoms shown by patients are also called 'presentation' or 'findings.' Doctors choose the right intervention for a disease based on the presentation of the patient. This process is called therapeutics.

Doctors make decisions other than diagnoses and therapeutics, and as such CDSS can be used to support a variety of decisions in a clinical setting. CDSS can provide alerts, reminders and recommendations for a variety of clinical decisions mostly related to the administration of drugs or medications (Khairat et al. 2018). In my study, I focus on CDSS that support diagnosis and therapeutics only. Such CDSS belong to the expert system category (defined and explained below). Even within diagnosis and therapeutics, any single CDSS cannot support all possible decisions related to all diseases. Hence, I focus on a subset of diseases that have skin manifestations, and I have used the term dermatological diseases for this subset. Dermatological diseases affect other body parts, organs and systems in addition to skin, as it is rare to find diseases that affect only one organ in the human body. I have used two relatively common dermatological diseases widely in this **book** as examples; erythema multiforme and psoriasis (both have no particular significance other than providing a context and example). Dermatologists are a group of doctors who have received special training to identify dermatological diseases.

1.3   Nature of knowledge in healthcare

An expert system requires a knowledgebase to support diagnostic decision support. Hence, I briefly introduce the nature of knowledge and the subset of knowledge that expert systems capture along with peculiarities of decision making in healthcare. These topics are reviewed in detail in a later section below. A doctor's knowledge may be factual (knowing a set of facts such as risk factors for psoriasis), conceptual knowledge (knowing the association and correlations such as psoriasis may be associated with scaly plaques on the elbow) and procedural knowledge (how to administer PUVA, a form of psoriasis treatment) (Patel et al. 2002). I mostly study the conceptual dimension of knowledge, and the word, *knowledge* (if not qualified further) refers to this conceptual dimension. Further, diagnosis is not a solitary process performed by the doctor, but a combined effort of several actors, including various IS. My examination of diagnosis as a process is rooted in the information integration theory (Anderson 1981).

I believe that a qualitative approach is needed to understand the complex healthcare context in which CDSS operate. CDSS are continual-learning systems, an important distinction that sets them apart from other HIS. CDSS are knowledge augmenting systems that cannot be segregated from the end-user for objective evaluation. The randomized controlled trials commonly used in the healthcare domain may not be suitable to adequately capture the effect of socio-technical systems such as CDSS in augmenting a clinician's decision-making abilities.

1.4 Structure

A detailed upfront review of literature is not recommended in GT because of the potential to introduce bias. I did heed this advice, and most of the literature review was done post hoc after analysis, though it appears next in this book. The literature review follows a logical structure in which I first describe the theoretical background and some of the existing IS subdomains, such as Decision Support Systems (DSS) and Knowledge Management Systems (KMS), that can provide a framework to investigate CDSS. I deconstruct common DSS concepts such as expert systems and recommender systems. Next, I describe human behaviour and the challenges related to decision-making in a healthcare context. Finally, I describe the challenges involved in CDSS design.

In the next section, I state my research questions and then proceed to describe the Charmaz (2006) method of GT and how I followed it in the methodology section. In the results section, I describe the systematic coding and sensemaking process, adopting the Gioia et al. (2013) method of analysis. I triangulate demographic and clickstream data and iteratively distill the codes into a substantive theory for explaining and predicting user adoption of CDSS systems based on the knowledge needs of the user. Then, I propose prescriptive guidelines for CDSS design.

In the discussion section, I discuss the insights I derive from my research questions and compare my findings to existing IS literature. Next, I describe the scope of my theories as recommended by Urquhart et al. (2010) and the typology according to the Gregor (2006). Then I state the limitations, some problems faced, and the implications of my findings. Guidelines for other researchers who wish to adopt my methodology of using a real stimulus are summarized.

Lastly, as an e-health hybrid, I believe that an IT artifact is at the centre of any IS inquiry, and its design cannot be ignored. The design of the stimulus is explained in detail in Appendix A, utilizing the novel and emerging concept of multi-modal machine learning and Resource Description Framework (RDF) for knowledge abstraction. In the same section, I describe the functional modules of the stimulus and how the stimulus design aligns with my sensitizing theory — the Theory of Planned Behaviour. To foreshadow my later discussion, having the same upfront theory for developing the interview guide and the artifact was very helpful in sensemaking. In addition to the theories to explain CDSS adoption and design, the main contributions of this study are the artifact design and using it as a real stimulus in a GT inquiry.

Some of the conventions that I have followed in this document are as follows. Quotes from the participants are italicized. Any text that I have added for clarity are in square brackets. Some of the concepts are elaborated in a glossary at the end of this document.

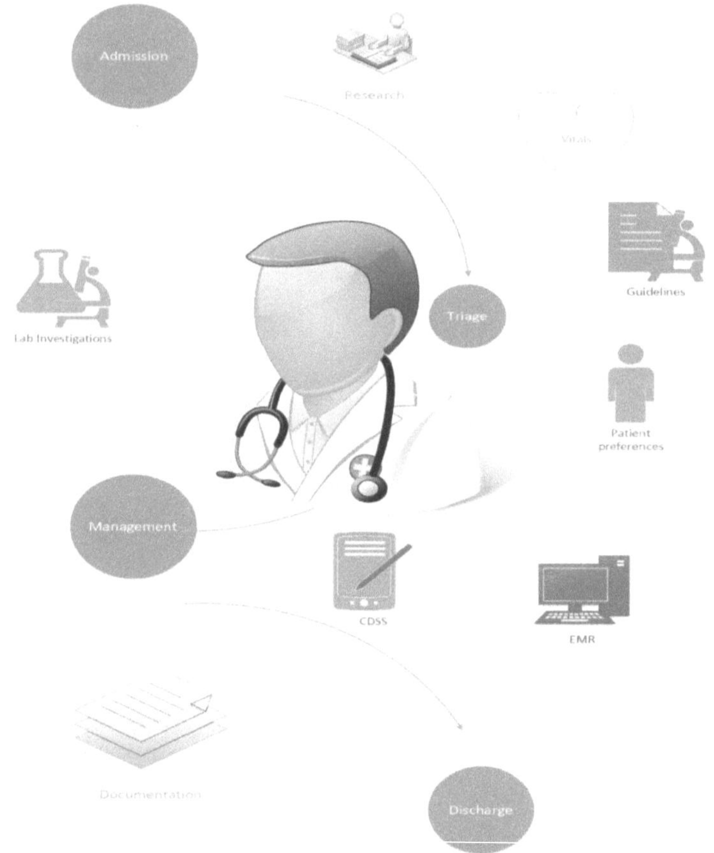

Figure 1.1: The complex context in an ER

Chapter 2

Review of literature

2.1 Introduction

It is not customary to use an extensive literature review in GT inquiry. This section aims to inform the readers of the existing knowledge concerning CDSS adoption and design. As the main objective of this study is to explore CDSS adoption by doctors, kernel IS theories on user adoption are discussed first with an emphasis on important constructs and their known relationships. Then I contextualize high-level IS theories to HIS in general and CDSS in particular by describing how user acceptance leads to two of its significant consequents: user adoption and system success. Next, CDSS is described within the context of two established subdomains in IS (DSS and KMS) to understand commonalities and differences. This is followed by the provision of an overarching taxonomy of CDSS, and a brief discussion of the state of the art in dermatology CDSS, the CDSS type that I use as a stimulus in this study.

GT aims to build theory as opposed to testing or validating theories. Theory building needs an understanding of core domains such as psychology and behavioural science. To align with this endeavour, I discuss the fundamental principles of Human-Computer Interaction (HCI) and decision science in the context of CDSS.

The use of a real CDSS as a stimulus is a unique feature of this study. Hence, I introduce the fundamental principles of Design Science Research (DSR) and the role of theory in the design process. This allows discussing previous attempts at combining GT with DSR, contributing both substantive theory for explaining behaviour and prescriptive design knowledge to guide artifact design. Finally, I conclude this section with a discussion of challenges in CDSS design. The research questions and scope are summarized in the next chapter.

2.2 Theories and subdomains

The nature and even the definition of theory vary according to the ontological and epistemological views of the domain. I follow Gregor (2006)'s (page 616) definition of theory in IS as *"abstract entities that aim to describe, explain, and enhance understanding of the world and, in some cases, to provide predictions of what will happen in the future and to give a basis for intervention and action."*

2.2.1 The typology of theories in IS

Gregor (2006) describes five types of IS theories as described below:

- Type 1: Theory for analysis with no causal relationships specified and predictions made.

- Type 2: Theories that provide explanations, but no predictions or testable propositions.

- Type 3: Theories with emphasis on accurate predictions and testable propositions, but no causal explanations.

- Type 4: Theories that provide predictions, testable hypotheses and causal explanations. Bhattacherjee and Premkumar (2004) proposed a theoretical model of user beliefs and attitudes toward IT usage that explains and predicts the change with time.

- Type 5: Theories that give explicit prescriptions for constructing an artifact. They describe how to 'do' something. The design theory for emergent knowledge processes providing prescriptive system principles for system development is an example (Markus et al. 2002).

The nature of theory depends on the domain of interest. Type 5 theories are common in DSR, while other types with well-defined constructs and falsifiable relationships are common in traditional IS research.

2.2.2 Theorizing in GT

GT aims to generate theory inductively grounding the account on data (Turner 2016) and, as such, is very useful for theorizing based on the context and processes associated with IS (Myers et al. 1997). Glaser and Strauss (1967) stresses the importance of theoretical coding by establishing the relationships between categories. During the process, narrow concepts undergo iterative conceptualization into substantive theories and finally into a formal theory

(Urquhart et al. 2010). Urquhart et al. (2010) suggests that GT is capable of generating any of the five types of theories in IS proposed by Gregor (2006). The level of conceptualization is also equally important as the scope in GT (Urquhart et al. 2010). Theoretical coding aims to generate inferential and predictive statements by proposing direct relationships between core categories. Analytical memos play a vital role in this process.

2.2.3 The artifact as theory

IS is a specialty that sits at the intersection of technology and behavioural sciences, and it is crucial to find an appropriate balance between science (theory) and technology (design artifact) (Baskerville et al. 2018). IT strategy should align with business strategy, and IT infrastructure should align with organizational infrastructure (Henderson and Venkatraman 1993). This applies to CDSS as well.

From a DSR perspective, an IT artifact can be categorized as a construct, model, method, or instantiation (March and Smith 1995). Constructs are the terminology and fundamental concepts; models are real-world problems represented using constructs; methods represent processes or algorithms; the instantiations demonstrate that all three can be implemented in a working system. An instantiation in DSR is equivalent to a theoretical contribution in behavioural IS (Baskerville et al. 2018).

2.2.4 User adoption

The attitude of a user towards any technology is one of the fundamental areas of inquiry in IS research, and as such, there are many theories to explain the psychology and behaviour associated with this. Bandura et al. (1999) and Ajzen

Figure 2.1: User acceptance consequents

and Fishbein (1980) set the stage for this inquiry with their pioneering works. The introduction of the popular Technology Acceptance Model (TAM) (Davis 1989; Davis et al. 1989) improved our understanding of Health IT adoption (Holden and Karsh 2010), and it evolved from two earlier theories in psychology; the Theory of Reasoned Action (TRA) (Ajzen and Fishbein 1980) and TPB (Ajken 1991). Table 1 summarizes some of the core constructs in these theories on user acceptance. User acceptance is seen by many as the antecedent of user adoption and, subsequently, system success (Figure 2.1).

Understanding user adoption and system use is at the core of IS research (Venkatesh et al. 2007). TAM is still one of the most widely used models in IS that describes *perceived usefulness* and *perceived ease of use* as two distinct antecedents of system use (Davis 1989; Davis et al. 1989). TAM draws from TRA from social psychology (Ajzen and Fishbein 1980). TRA was later expanded to TPB (Ajken 1991), the sensitizing theory used in this study. To set the stage for TPB, I briefly explain these models and their constructs along with some of the later models that emerged that further qualify them. Table 2.1 summarizes the constructs and theories in IS related to user adoption.

Table 2.1: Popular theories and constructs in IS related to user adoption

Constructs	Theories
Perceived usefulness (PU)	TAM (Davis 1989)
Perceived ease of use (PEOU)	TAM (Davis 1989)
Attitude (AT)	TAM (Davis 1989)
	TRA (Ajzen & Fishbein 1980)
	TPB (Ajzen 1985)
Self-efficacy (SE)	TPB (Ajzen 1985)
Social norms (SN)	TPB (Ajzen 1985)
	TRA (Ajzen & Fishbein 1980)
Trust	UTAUT (Venkatesh et al. 2003)
Behavioural intention (BI)	TAM (Davis 1989)

Table 2.1 Continued: Popular theories and constructs in IS related to user
adoption

System use (SU)	DeLone and McLean model (Delone and McLean 2003)	
System quality (SQ)	DeLone and McLean model (Delone and McLean 2003)	
Information quality (IQ)	DeLone and McLean model (Delone and McLean 2003)	

Theory of reasoned action (TRA)

TRA postulates that the person's intention to perform a behaviour is the best
predictor of that behaviour (Ajzen and Fishbein 1980). Antecedents of intention
are the attitude towards the behaviour and the perceived expectations of
significant others concerning the individual performing the behaviour.

Technology acceptance model (TAM)

TAM introduced two constructs that formed the basis for much IS research:
Perceived Usefulness (PU) and Perceived Ease-Of-Use (PEOU). PU is defined as
the "the degree to which a person believes that using a particular system would
enhance his or her job performance," and PEOU is defined as "the degree to
which a person believes that using a particular system would be free from effort"

(Davis et al. 1989). Both are postulated to have a positive effect on the behavioural intention to use, an antecedent for the actual system use. The TAM model has been revised and modified several times with the addition of more constructs. Among them, TAM2(Venkatesh and Davis 2000) and the Unified Theory of Acceptance and Use of Technology (UTAUT) (Venkatesh et al. 2003) are the most prominent. TAM2 introduced a few more constructs (subjective norms, voluntariness, image) to the original TAM model to represent social influence. UTAUT is a comprehensive theory that integrates eight previous theories of user acceptance. One of the main contributions of UTAUT is the introduction of trust as an important construct.

TAM and the other theories have been used in the context of CDSS to explain user adoption (Schaik et al. 2004; Wendt et al. 2000). In general, performance expectancy, a construct in UTAUT, was found to be a strong predictor of user adoption. Johnson et al. (2014) demonstrated the positive impact of PU and PEOU. Shibl et al. (2013) confirmed trust as an essential construct in CDSS adoption. Some of the important factors with an unfavourable impact were workflow interference, questionable validity of the systems, and excessive disturbances such as alerts (Khairat et al. 2018). The correlation between these constructs and patient outcome is rarely studied, and whether system success leads to improved patient outcomes needs to be established as well (Wallace et al. 1995). However, there is some consensus that suggestions offered by CDSS may be useful to a patient (Sousa et al. 2015; Zheng et al. 2005).

DeLone and McLean model

User adoption in itself cannot ensure system success, though it is a necessary procondition (Petter et al. 2008). The updated DeLone and McLean model is a popular model in IS to predict success based on six constructs; system quality, information quality, service quality, use, user satisfaction, and net benefits (Delone and McLean 2003). Some of these constructs need consideration in the context of CDSS. System use may not be an indicator of system success if the use is mandated, as in most HIS, including CDSS (Seddon and Kiew 1996). In such cases Seddon and Kiew (1996) argues that PU from TAM would be a better indicator of success.

I describe some of the success dimensions from the DeLone and McLean model because of its significance in my results. System quality is the desirable characters of an IS such as ease of use, ease of learning and response times. Information quality is the desirable features of the output. System use includes amount, frequency, appropriateness, nature and purpose of use (Petter et al. 2008). Net benefits in the context of CDSS applies to the improvement in patient outcomes. Another important implication of this model is the need to measure all six dimensions of success for evaluation(Petter et al. 2008). Most CDSS evaluations focus on the accuracy aspect of information quality. Sedera et al. (2004) proposes a validated instrument for measuring system success. However, a generic IS success evaluation instrument may not be relevant in the context of CDSS.

Diffusion of Innovation (DOI) theory

DOI is a theory proposed by Rogers to explain how an idea or product gains momentum and diffuses (or spreads) through a specific population or social system (Rogers 1981). DOI specifies five adopter categories: innovators, early adopters, early majority, late majority and laggards. A small fraction of the population are innovators who want to try the innovation at any cost. A few users are early adopters who embrace change, and the rest take varying amounts of time to adopt. Factors such as relative advantage, complexity, compatibility, trialability and observability influence adoption. DOI has been used to explain the adoption of clinical change (Sanson-Fisher 2004). Willingness to use a new intervention in the medical domain is influenced by, specialty, medical school, experience, practice location and practice volume (Tamblyn et al. 2003).

The existing theories of technology adoption have been unsuccessful in describing HIS adoption in general and CDSS adoption in particular. Holbrook summarizes that "factors that would predict successful CDSS still have not been adequately identified"(Holbrook et al. 2003).

Next, I discuss TPB – the up-front theory that guided GT and design – in detail.

2.2.5   Theory of planned behaviour (TPB)

The theory of planned behaviour was proposed by Ajzen (1985) suggesting that attitude toward behaviour, subjective norms, and perceived behavioural control, together shape an individual's behavioural intentions and subsequent actions. It is an extension of TRA. Subjective norm represents behaviour in the presence of perceived social pressures from significant referents and the actors' motivation to comply with the referents. Social media interactions may be significant in defining and modifying social norms (Kiriakidis 2017).

TPB extends TRA (Fishbein 1979) by introducing a new construct, Perceived Behavioural Control (PBC), defined as "an individual's perceived ease or difficulty of performing... [a] particular behaviour" (Terry and O'Leary 1995). PBC is related to self-efficacy, but it is not as extensively studied in IS as self-efficacy. PBC is assumed to reflect past experiences and anticipated obstacles (Ajken 1991). The resources available to the clinician at the point of care may influence PBC in the context of CDSS.

TPB provides us with a theoretical lens to understand knowledge augmenting systems in a resource-deprived region (Kiene et al. 2014). A clinician practicing in a resource-deprived area is likely to have experienced significant obstacles related to knowledge acquisition from peers and other external knowledge sources such as published biomedical literature at the point of patient care. A knowledge augmenting clinical decision support system that is easily accessible is likely to increase the PBC of clinicians. Attitude towards behaviour in TPB is related to outcome expectancy (Williams 2010). CDSS may improve outcome expectancy when other knowledge resources are not available, leading to better adoption intention.

2.2.6 Decision support systems (DSS)

DSS is an important area of IS that emerged in the 1970s with a focus on IT systems that support managerial decision making in the context of firms (Arnott and Pervan 2008). DSS is the overarching domain with a number of sub-divisions for research and practice such as personal DSS, group support systems, intelligent DSS, knowledge management-based DSS, business intelligence and data warehousing (Arnott and Pervan 2005). DSS concepts and subgroups are described mostly in the context of firms and, as such, may not be directly applicable to CDSS. CDSS have unique needs, such as patient safety and privacy (Fichman et al. 2011), and the various stakeholders may have divergent interests. CDSS are non-homogenous and can belong to multiple subgroups. Next, I will briefly introduce some of the common characteristics shared by DSS and then explain how CDSS differs from DSS in other domains.

Wicked problems are complex problems that are hard to solve using traditional methods because of incomplete knowledge and multiple stakeholders (Pretorius 2016). Such problems are common in healthcare (Raisio 2009). DSS assist in storing, retrieving, and manipulating information to arrive at a decision in such complex problems. DSS, by definition, are adaptable to changes in the environment and the decision making approach (Sprague 1980). Some of the issues faced by DSS are the widening gap between research and practice, lack of established research methods and paradigms, emphasis on technology rather than theory, and a lack of coherence in the discipline (Arnott and Pervan 2008). Many of these problems are shared by CDSS as well.

CDSS are different from other DSS in the following ways (Fichman et al. 2011):

1. Patient safety is of vital importance in CDSS, and the stakes of making mistakes are very high and can potentially lead to loss of lives. Consumers of healthcare services believe that healthcare experts could not and should not fail (Patel et al. 2011). The errors related to CDSS can be procedural (following a wrong process) or interpretive (following wrong information or the wrong interpretation of information). Training combined with CDSS

mediated automation can potentially reduce errors but, unfortunately, may
increase some errors (Aron et al. 2011). The need and the inadequacy of
existing CDSS mediated automation, and the need for a better approach
became painfully apparent during the recent COVID-19 pandemic.

2. Patient privacy is important and can be highly regulated.

3. Healthcare is hierarchical and professionally driven and, as such, has a
   significant impact on adoption.

4. Healthcare is multidisciplinary, with each team having a variety of skills
   and functions. Healthcare professionals in varied fields with different
   backgrounds and experience should cooperate seamlessly for effective
   patient care (Laxmisan et al. 2005). Better patient outcomes need a great
   deal of coordination between teams. Besides, the number of stakeholders
   and their priorities are many, with an attempt towards *unity in diversity*
   (Oborn et al. 2011).

5. CDSS implementation is complex and should adapt seamlessly with the
   changing needs and requirements.

As the type of CDSS that I explore manage everchanging healthcare knowledge
and evidence, I explore some of the KMS paradigms that are important in my
exploration.

2.2.7 CDSS as knowledge management systems

Knowledge Management deals with the creation, storage, retrieval, transmission
and application of knowledge (Alavi and Leidner 2001). Clinical knowledge is

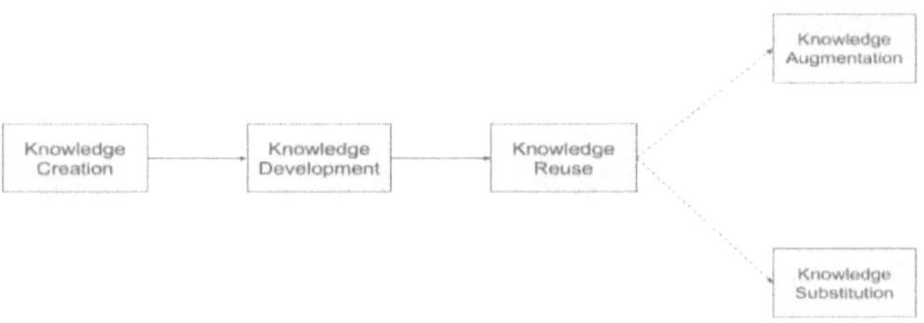

Figure 2.2: The knowledge value chain

difficult to collect as it continually evolves as new evidence is generated from clinical trials. Clinical domain knowledge requires deconstruction into a general form (decontextualization) and reconstructing it in the context of the given patient (recontextualization) (Friberg et al. 2000). Clinical knowledge is usually compiled by a group of specialists and is captured in published literature. The quality of captured clinical knowledge tends to be generally low (ambiguity exists) with a moderate life span of knowledge (new evidence emerges) (Malterud 2001). The decision space of CDSS may be focal (confined to a few related diseases or conditions), specialty-focused (limited to a single medical specialty) or generic (Karargyris et al. 2012).

Knowledge reuse is the practical application of the knowledge captured in an IS (Dennis and Vessey 2005). The two main knowledge reuse goals are knowledge substitution and knowledge augmentation (see Figure 2.2). Highly structured knowledge repositories substitute the knowledge of the end-user. The end-user has limited prior knowledge of the contents of the knowledge repository. Knowledge augmentation systems augment end-user existing knowledge with additional information that needs contextualization. Clinical practitioners augment their existing knowledge by reusing each other's knowledge contributions (Miller 1990). CDSS are knowledge augmentation systems (Dennis and Vessey 2005), and their use depends heavily on the existing knowledge of domain experts.

2.2.8 Expert Systems

Expert systems are domain-specific DSS that rely on a domain-specific knowledgebase and a general inference engine (Myers 1986). Expert systems can be useful in disseminating knowledge, relieving doctors from routine tasks and conveniently testing ideas and theories (Thomas 1989). Expert systems can efficiently perform some of the repetitive tasks and relieve doctors for more complex tasks. Expert systems can replace human experts or aid them. Despite their advantages, such as availability and comprehensiveness, the uncertainty associated with knowledgebase validity and user responses is a limitation (Metaxiotis and Samouilidis 2000). Expert systems are beneficial in diagnostics, therapeutics and medical education. One of the earliest and the most popular expert systems in medicine is Mycin, developed at Stanford University (Shortliffe 1977).

2.2.9 Definition and taxonomies of CDSS

I will use the following working definition of CDSS: CDSS refers to systems providing clinicians with computer-generated clinical knowledge, intelligently filtered to enhance patient care.(Osheroff et al. 2007)

CDSS can be classified according to the capabilities they provide as follows: (Garg et al. 2005)

- Alerts and reminders

- Modelling and prediction tools such as various clinical calculators

- Those that focus on information retrieval (Example: InfoButtons (Del Fiol et al. 2012))

- Those that support decision making and,

- Those that support complex clinical workflows.

CDSS can also be classified according to the type of clinical decision process they support, into diagnostic and therapeutic CDSS (Sutton et al. 2020). The decision-making in healthcare has many distinct types of logical inferences such as clinical goals, treatment options, identifying relevant evidence, tailoring evidence to the given patient and assessing the overall merit of a decision (Fox et al. 2010).

2.2.10 CDSS in Dermatology

I briefly review state of the art in dermatology CDSS as the stimulus used in this study is restricted to dermatology. Dermatology – the medical specialty that deals with skin diseases – adopted Machine Learning (ML) and Artificial Intelligence

(AI) based CDSS early on. Most CDSS in dermatology aim to identify melanoma – a lethal skin cancer – early in its course so that the patients can be effectively managed and lives saved. These melanoma screening CDSS, usually available as patient-facing mobile apps, have been an enormous success, and in some cases, they have been shown to outperform trained dermatologists (Brinker et al. 2019). Because of their popularity, CDSS in dermatology is almost synonymous with melanoma screening apps.

Dermatologists make decisions based on what the patient says (clinical history) and what is seen (physical examination). This requires a multi-modal approach, and dermatology CDSS need to be trained on both text and images. The training of ML-based CDSS has some unique challenges, that require specialized solutions. (Eapen et al. 2020).

The skin, being the largest organ in the human body, accounts for a substantial proportion of ailments reported by patients, and as such, the number of dermatologists needed to manage them is less than required in many countries, including Canada (Curran et al. 2007). Many of the patients with skin ailments are managed by primary care physicians, and recently there has been a call in Canada for supporting them with AI tools (Lewinson and Vallerand 2020).

2.3   Human behaviour

2.3.1   Human-computer interactions

HCI plays a vital role in CDSS design, and with the advances in ML and AI-based CDSS, HCI requires a closer look. In order to ensure patient safety, it is vital to combine AI software and human cognition capabilities by taking into account safe interaction paradigms between the two (Topol 2019). Explainability and interpretability play a significant role in the successful deployment of CDSS (Vellido 2019).

Graphical User Interfaces (GUI) are a major cause of concern for doctors, especially when designers are more technology-oriented than design-oriented (Rundo et al. 2020). GUI needs to be tailored to the clinical context taking into account cognitive aspects. A thorough usability testing of a GUI using methods such as the read-aloud method is necessary. Natural Language Processing (NLP) shows tremendous potential in improving GUIs (Esteva et al. 2019).

User-Centred Design (UCD) is emerging as an important paradigm in HCI. It is an iterative design process with a focus on the users and their needs. There are several instances of successful UCD in healthcare (Brennan et al. 2010), including a novel cognitive approach (Malhotra et al. 2005).

HCI studies have demonstrated an increasing cognitive overload for doctors. King et al. (2019) demonstrated the importance of selectively highlighting critical patient information in decision-making. Systems that are context-aware, providing case-based reasoning to doctors, show promise (Rundo et al. 2020). The effective use of internet-based shared resources such as PubMed has been recognized. An emerging trend is to include patient preferences in CDSS to realize a shared decision making process (Peleg et al. 2017).

Explanations provided by CDSS

CDSS should provide users with an adequate explanation about the choices recommended by the system in order to ensure user trust. The type, frequency and content of the information provided by CDSS are important in its adoption. The family of advice-giving IS is called recommender systems (Nunes and Jannach 2017). Their explanations may pertain to the decision choice provided or the alternatives offered. Some recommender systems provide information on peer preferences and the data and rationale for the inference method. How these explanations are displayed (e.g., adaptive or automated) and the level of detail of the explanations are important. The display can be in the form of canned text, lists, template-based or structured representations (Hwang et al. 2020). Explanations for decision choices are becoming increasingly important in the era of 'black box' ML and AI systems (Wang et al. 2019).

2.3.2 Clinical decision making

With the increasing importance of CDSS in clinical decision making, there is a need for re-conceptualizing the old design foundation to adapt it to the new digital world. It is especially important as the stakes of bad decisions can be as high as loss of human life (Patel et al. 2002). Classical decision theory uses Bayesian and probability theories to choose the optimal option from an array of options that maximizes a utility function (Beach and Lipshitz 2015). However, in the clinical context, decision making is not a wholly computational or analytical process but has an equally important affective component (Bechara et al. 1997). Decisions made under natural circumstances associated with clinical decision making such as stress, time pressure and limited resources may not conform with the notion of optimal decisions guided by Bayesian probabilities (Heckerman and Shortliffe 1992).

Decisions carry different meanings and have different consequences to various healthcare stakeholders (Medin and Bazerman 1999). In many cases, technology fundamentally transforms the decision making process (Patel et al. 2002).

The traditional decision science view is that all rational actors try to maximize 'subjective expected utility' and any systematic deviation from this is construed as bias. The number of variables in clinical decision making is large, and it is impossible to choose the correct ones and ignore the rest. Any error in choosing the right variables leads to an error in judgement (Baron 2000). CDSS may reduce the cognitive burden in choosing the right decision variables, though such CDSS may not directly influence the ultimate decision outcome.

Doctors are also prone to some well known psychological biases (Chapman and Elstein 2000). Recall bias refers to the tendency to recall a vivid encounter with a patient, leading to inappropriate diagnosis in another patient with a similar presentation. This is related to availability bias that occurs when physicians diagnose cases that have similarities with recently encountered cases (Mamede et al. 2010). Doctors (like other individuals) are prone to anchoring (giving excessive weight to initial information), inability to correctly interpret statistical cues, the bias in evaluating probability and hindsight bias (tendency to overestimate the ability to make a correct diagnosis after the fact). The framing bias refers to changes in choices, depending on how the problem is described or framed. All these biases can lead to suboptimal diagnosis and subsequent management (Croskerry 2003). CDSS could play a role in mitigating some of these biases.

Traditional decision science research in the context of healthcare is subsumed by two distinct traditions (Patel et al. 2013): i) a problem-solving tradition that emphasizes cognitive processes and takes expert decisions as the gold standard;

and ii) a decision-analytic tradition that sees decisions as rational choices under uncertainty with experts perceived as fallible agents liable to biases (Patel et al. 2001).

Doctors arrive at an initial hypothesis based on the findings of the patient, a process similar to classification in ML (Patel et al. 2002). Research shows that once doctors arrive at a hypothesis, subsequent mental processing is directed towards confirming the hypothesis (hypothesis evaluation) (Evans and Patel 1989). This tendency is more prominent in senior doctors than residents and can occasionally lead to confirmation bias. This may have an impact on the way senior doctors and residents use CDSS. The paradigm of naturalistic decision making is vital in this context, as cognition in real-world clinical environments may be significantly different from most other stable environments (Klein et al. 1989).

Unlike most other domains that rely on bounded rationality (simple heuristics) for decision making, healthcare depends on ecological rationality that takes the fluid environmental conditions into account (Forster 1999). In clinical decision making, the ability to reason and act are dissipated among several actors throughout the environment to reduce the cognitive load on any one individual (Khairat et al. 2018). Doctors need mental models that represent the dynamic representations of the fluid environment, each situation being unique (Holland et al. 1989).

Finally, the distinction between *effects with* and *effects of* technology is important in the context of CDSS (Salomon et al. 1991). The availability of CDSS gives doctors the ability to gather information more systematically and efficiently. The *effects with* CDSS reduce end-user cognitive load and allow these systems to focus on hypothesis generation and evaluation. CDSS can also introduce long-lasting changes in the decision-making process. For example, a CDSS that analyzes recent articles for diagnosis may encourage a doctor to follow the same method even when the CDSS is not available. These *effects of CDSS* may improve the knowledge and skills of the doctor.

2.3.3   External factors

User adoption of CDSS may be affected by factors other than clinical or technological considerations. Limitation of autonomy may lead to inadequate or inappropriate system use (Walter and Lopez 2008). Doctors are prone to reactance – "a motivational state whereby people react to situations to retain freedom and autonomy" (Vashitz et al. 2009).

2.3.4   Adoption of CDSS

Healthcare is a highly knowledge-intensive industry with several classes of stakeholders and beneficiaries. Existing IS and knowledge management theories do not explain or guide CDSS adoption in health settings (Carbone 2010). This is partly due to an obsession with a technology-driven implementation and a focus on business outcomes from IS researchers (Aarts et al. 2004; Bates et al. 2001; Chaudhry et al. 2006). In contrast, clinical practitioner teams prioritize evidence-based medicine (EBM) and health outcomes (Heckley 2004).

Several factors hinder the take-up of CDSS by clinicians(Greenes 2011), such as a lack of standards (Tu et al. 2007) and a lack of formal knowledge representation methods that are scalable, reusable and interoperable (Fox et al. 2010). There is a need to create "learning health systems" that can periodically update the knowledge base (Cresswell et al. 2016; Henry et al. 2016).

CDSS designers should strive to support as many processes in clinical decision making as possible, such as diagnosis, workflow management and care planning (Fox et al. 2010). CDSS implementation failures in the healthcare setting are common (Dwivedi et al. 2011; Wade et al. 2006). However, it is difficult to objectively define the failure and success of CDSS as there are many contexts, stakeholders and beneficiaries. Defining the success and failures of CDSS is not an objective of this study. However, a holistic understanding of success and failure is essential for appreciating the phenomenon of the adoption of CDSS (Heeks 2006). The failure (and success) of CDSS can be total or partial depending on the degree of fulfilment of stakeholder objectives. As there are multiple stakeholders, the objectives are complimentary at best and divergent at worst. For example, a cancer patient may prioritize the quality of life over longevity while the patient's doctor may have different priorities. Hence an objective evaluation of CDSS is not possible in most cases (Friedman and Wyatt 2013). To improve the evaluation of CDSS, qualitative evaluation methods that use an iterative approach should be used to complement quantitative methods in providing insights into the success and failure of CDSS (Kaplan 2001).

Heeks (Heeks 2006) recommends the design-reality gap framework for understanding the success and failure of CDSS. The design-reality gap framework posits ITPOSMO dimensions that can potentially influence success. ITPOSMO is an acronym for information, technology, processes, objectives and values, staffing, management systems and other resources.

There may be a mismatch between what information a systems provides and the information needs of the clinician (Chiasson and Davidson 2004). For example, a system may provide statistics on skin cancer, but this may not be important for a dermatologist when assessing a patient. The non-availability of technical requirements is a limitation in some healthcare settings (Teixeira et al. 2012). For example, optimized virtual machines for machine learning in an image analytics system may not be available in a hospital.

Epistemological concerns exist in the design of randomized control trials (Moher et al. 1995), conducting systematic reviews (Sacks et al. 1996) and creating disease management guidelines (Shaneyfelt 2012). Clinical practitioners do not always rely on literature evidence (Ellis et al. 1995; Nordin-Johansson and Asplund 2000). The movement towards patient-centric healthcare puts patient values and preferences at the centre of clinical practice. Hence, CDSS systems that support clinical practice should not be just algorithms that recommend treatment based on statistics alone. The evaluation of the correctness, reliability and validity of the CDSS knowledge base is necessary but not sufficient (Sim et al. 2001). The objectives and values of practitioners in resource-deprived areas may contrast with that of others. Adequate staffing and sufficient staff skills are important since designing a care plan requires considerable cognitive resources and time (Fox et al. 2010).

Successful CDSS design and implementation requires a combination of good design theories and insights from the practical experience of domain experts (Fox et al. 2010). The nature of the clinical decision-making process is different (Goldstein and Hogarth 1997; Medin et al. 1995) from other domains, and an understanding of the nature of clinical judgment is important in ensuring patient safety (Dowie and Elstein 1988; Schwartz and Griffin 2012).

2.4   Design science research (DSR) in IS

DSR is a prominent and distinct research paradigm in IS (Iivari 2007), and it promotes the view that IT artifacts play a central role in IS research. The research contribution from a DSR can be constructs, models, methods, or instantiations, or a combination thereof (March and Smith 1995). Traditionally, IT artifact refers to a tangible instantiation. DSR is iterative by nature (Baskerville et al. 2009) and it deals with real solutions to real-world problems. DSR is different from routine design in producing generalizable and novel design knowledge in addition to an IT artifact. The role of theory is frequently debated in DSR. They can be kernel theories, design relevant explanatory/predictive theories (DREPTs) and Information System Design Theories (ISDTs) (Kuechler and Vaishnavi 2012). Next, I briefly discuss the role of DREPTs and compare it with the other two, because of its importance in this study.

2.4.1   Design relevant explanatory/predictive theories

High-level kernel theories in IS, such as TAM, cannot directly guide design. Kuechler and Vaishnavi (2012) describes two types of design theories. DREPTs are mid-range theories that map constructs in kernel theory to design features. ISDTs are low-level design knowledge that directly guide artifact design. In the hierarchy of IS theories for design, DREPTs retain a certain level of abstraction compared to ISDTs but are not as abstract as kernel theory. Within the framework of Gregor (2006)'s typology of IS theories, Kuechler and Vaishnavi (2012) position ISDTs as type 5 theories for design and action, and DREPTs as type II-IV (explanatory/predictive) theories with design implications. In short,

DREPTs explain how or why artifact features have the desired effect while ISDT describes how to construct an artifact (Kuechler and Vaishnavi 2012). Using GT to guide design is tricky because of epistemological contradictions (Holmström et al. 2009). However, it remains crucial to "generate theories that work" (Manning 1992). GT and DSR can potentially complement each other (Holmström et al. 2009), and Goldkuhl (2004) proposes prescriptive guidelines for using GT in a design context.

Next, I explain the distinction between the notion of generalizability, projectability and contextualization in the context of GT.

Generalizability

The notion of generalizability is mostly viewed as a positivist concept in a quantitative tradition (Carminati 2018; Delmar 2010; Kerlinger and Lee 1999). Generalizability is a vital requirement in healthcare where evidence-based knowledge is generated from systematic reviews of published literature (Bradt 2009). As most of the participants of this study and potential readers are likely to be from the healthcare domain, I reiterate the difference in generalizability in this GT inquiry from traditional randomized controlled trials in healthcare based on randomly selected samples representative of the population (Lee and Baskerville 2003). The generalization in GT is achieved by following a rigorous theory-building process that involves coding, constant comparison and conceptualization till theoretical saturation is reached (Glaser and Strauss 1967). The analytical generalization helps apply the results in other related contexts though it is not based on random samples from defined populations (Hallberg 2013).

Projectability

DSR generates prescriptive knowledge, and a notion of retrospective generalization may not be applicable in most contexts. In other words, generalization "looks backwards," while DSR describes future use. Baskerville and Pries-Heje (2019) introduced the concept of projectability in DSR as an alternative to generalization. Projection refers to instances that support a theory, and projectability refers to the ability to support relevant projections.

Contextualization

Qualitative inquiry methods with an interpretivist mindset may not be truly generalizable as in most quantitative traditions. Context is an important aspect of qualitative research, and as such, any research output should be contextualized and understood in terms of its context (Yilmaz 2013). Interdisciplinary fields such as IS may approach problems to produce knowledge that is ready to implement, and the notion of causality and universal generalizability should be approached with caution as the causality assumptions may be linked to the context (Andersen and Risør 2014).

2.4.2 User-centred design and prototyping

The importance of involving doctors early on in the design of any HIS is widely recognized (Kushniruk and Nøhr 2016; Tang et al. 2018). In user-centred design, designers take on the perspective of users through usability testing and user observation. In co-design, the designer works closely with end-users. The end-user is an active participant in the entire design process in participatory design (Kushniruk and Nøhr 2016). Despite its known benefits, it is a challenging goal to achieve because of the social, cultural, organizational and technical factors (Pilemalm and Timpka 2008).

A joint inquiry is a process through which a group of diverse individuals (doctors and software developers) who face an uncertain situation (effective CDSS design) jointly define and explore a problem, and jointly generate and evaluate different hypotheses about how to solve it (Avdiji et al. 2020). To communicate their ideas, software engineers generate prototypes that represent the solution (Avdiji et al. 2020).

2.4.3   DSR methodology

In DSR, knowledge is gained during the building of an IT artifact. Hevner et al. (2004) provides a framework and set of guidelines for understanding, executing, and evaluating DSR. The guidelines are summarized below.

1. The DSR output should be a construct, model, method or an instantiation (March and Smith 1995). The relationship between these levels are hierarchical in nature as shown in Figure 2.3.

2. DSR should solve a real-world problem. The problem space in DSR ideally should be generalizable.

3. The design should be evaluated appropriately

4. DSR should provide contributions in the areas of design artifact and design methodologies.

5. The DSR artifact construction and evaluation must be rigorous.

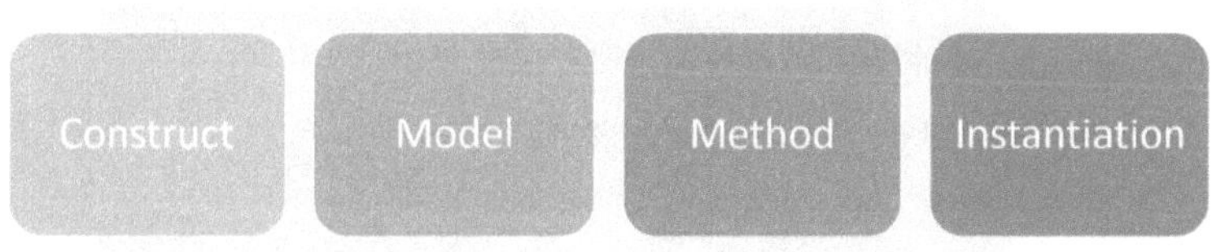

Figure 2.3: The nature of research contributions in a DSR project.

6. The artifact should use utilizable available means. There should be no attempt at reinventing the wheel.

7. The artifact must be communicated appropriately to the business-oriented and technology-oriented stakeholders.

2.4.4 Challenges in the design of CDSS

Sittig et al. (2008) identified the top ten grand challenges in clinical decision support using an iterative consensus-building mechanism. These grand challenges in CDSS design are summarized below.

It is crucial to improve the human-computer interface to support clinical workflow. Excessive alerts and reminders could lead to the problem of alert fatigue (Ash et al. 2007). Patient-level information should be summarized so that clinicians have immediate and direct access to the essential facts about the current patient. It is vital to prioritize recommendations to the user, rank recommendations according to the given patient and combine recommendations for patients with co-morbidities. Using text mining techniques to drive clinical decision support is vital as information is captured as free text in EMR systems and in published literature. Expertise is required for content development, implementation and to mine large clinical databases. It is vital for CDSS designers to disseminate best practices in design and implementation and to create and adopt standard architectures for sharing executable CDSS modules. Sittig et al. (2008) conclude by stating the importance of CDSS being accessible through the internet.

The evaluation of architectures for CDSS should focus on defining a set of desirable features, building a prototype and demonstrating that the architecture is useful. Some desirable features have been identified, such as an established ontology, shareable and viewable content integrated into the workflow with automated central updates (Wright and Sittig 2008).

2.4.5 Evidence-based medicine and CDSS

Evidence-Based Medicine (EBM) "is the use of modern, best evidence in making decisions about the care of individual patients by integrating clinical experience and patient values with the best available research information" (Masic et al. 2008). EBM requires efficient literature-search and evaluating available evidence at the point of care. CDSS as KMS play a critical role in identifying, creating, storing, retrieving and sharing evidence. Sim et al. (2001) recommend sharable, machine-readable up-to-date knowledge repositories integrated with electronic health records. Such Knowledge repositories should be rooted in formal ontologies with automated methods for updating knowledge. Automated systems are required to aid clinicians to keep up with recent evidence (Sim et al.

Sim et al. (2001) introduced the concept of evidence-adaptive CDSS that use a knowledge base that can reflect the most up-to-date evidence from literature and practice. Though published literature is the primary source of curated evidence, practice-based evidence from a clinician's experience should complement it. Specialized systems are required to capture such practical knowledge. However, patients trust clinicians in making the final decision regarding what is best for a patient (Hesse et al. 2005).

The impact of CDSS on clinical outcomes remains uncertain (Hunt et al. 1998), and some automation attempts may degrade the overall performance of CDSS (Woods and Patterson 2001).

2.4.6   Stimulus Design

A stimulus is an item used in behavioural research to elicit a response from the participants (Rozeboom 1960). DermML, the stimulus in this study (See Appendix A for details), supports clinical reasoning. Clinical reasoning cannot be bounded by "IF (condition)...THEN (action)" heuristics. It is challenging to incorporate the complexities of the context in which doctors operate as input into a CDSS. Further, it is difficult to externalize and capture the tacit knowledge residing as mental models in a doctor's mind. Additionally, the regulatory and privacy issues associated with healthcare make doctors view CDSS with suspicion. Any suggestion offered by CDSS must be sufficiently transparent and based on the best possible evidence (Khairat et al. 2018). This is a crucial consideration for ML and AI-based CDSS designers. If not appropriately designed with the context in mind, CDSS implementations can potentially harm the patients and lead to loss of lives (Harrison et al. 2007).

Chapter 3

Research questions and scope

3.1 Introduction

This study adopts a constructivist GT approach (Charmaz 2006) to investigate the gaps that exist in our understanding of the substantive area of user adoption of CDSS. In the previous chapter, I introduced the current state of IS research related to user adoption and the challenges in directly adopting them to CDSS. In this chapter, I discuss my research questions and explain the lens through which I, as a researcher defined the scope of research prior to data collection. However, pre-existing theoretical concepts did not confine or restrain interpretation of the collected data.

Sarker et al. (2013) describes popular patterns observed in qualitative research and provide guidance for researchers in developing qualitative research projects. Aligning with Sarker et al. (2013)'s recommendations, I discuss my research questions, theory, methodology and the nature of contributions and presentation. The design theory for stimulus design is described in Appendix A.

3.2   Research questions

CDSS are critical tools for doctors to provide better care efficiently, and yet their adoption is suboptimal. To understand this phenomenon, I ask the overarching question: how do doctors describe and characterize their use of CDSS in their practices?

I investigate the following aspects of CDSS adoption:

What are the characteristics of individual CDSS users?

CDSS may be more acceptable to physicians with certain learning styles, such as assimilating and converging style (Bergman and Fors 2005). Familiarity with ICT was identified as a critical factor in CDSS adoption in another study (Gagnon et al. 2012). It is widely believed that professional seniority is a barrier for adoption (Abdekhoda et al. 2015; Or et al. 2014). I wanted to explore individual characteristics such as age, gender, and familiarity with other CDSS.

How do doctors use CDSS in practice?

Prescription patterns of physicians are influenced more by CDSS than academic detailing (Buising et al. 2008). CDSS can be used as a prediction tool or for information retrieval. CDSS can also provide alerts and reminders (Garg et al. 2005). There is also the risk of over-reliance on potentially wrong suggestions provided by CDSS (Buising et al. 2008). I wanted to investigate whether doctors use CDSS to improve adherence to clinical guidelines, improve efficiency, administrative functions such as ordering tests, or diagnostic decision-making (Sutton et al. 2020).

Does the scarcity of resources impact CDSS use?

The sensitizing theory (TPB) suggests the possibility of the scarcity of resources having an impact on CDSS adoption and use. Currently, user adoption of CDSS in developing countries has not been adequately studied (Ahlan and Ahmad 2014). Not much is known about how CDSS impact the user in terms of "systems use, the effect on work processes, and whether it facilitates appropriate decisions by clinicians and patients" (Medlock et al. 2016). This knowledge gap may be more striking in resource-deprived areas.

What are the factors associated with enhanced CDSS use?

Computer literacy has been associated with enhanced CDSS use. Various user, environmental and organizational factors may influence CDSS use and adoption (Mah et al. 2020). Doctors' reimbursement is increasingly getting tied to process and clinical outcomes (Murphy 2014). CDSS may play a crucial role in improving patient outcomes (Garg et al. 2005) and this also might have an effect on its enhanced use.

Next, I ask a design related question:

What are the design factors that affect CDSS use?

Sittig et al. (2008) identified the grand challenges in CDSS design using an iterative, consensus-building process. Design issues have been identified as the most critical factor influencing CDSS adoption (Carroll et al. 2002).

3.3 The IT artifact

The presence of an IT artifact separates IS research from other research disciplines in business (Benbasat and Zmud 2003). Qualitative studies in IS often end up treating IT as an "omitted variable," and Orlikowski and Iacono (2001) issued a call for "theorizing the IT artifact." Sarker et al. (2013) also calls for a conscious attempt to include important technology-related topics in qualitative inquiry. This study has an IT artifact, firmly at the centre.

Putting an IT artifact at the centre of a qualitative study needs some careful considerations. As previously described, an IT artifact can be a construct, model, method or an instantiation (March and Smith 1995). From theory (Lukyanenko et al. 2015) and practice, I realized that instantiation is needed for the qualitative inquiry involving participants from non-IT areas – healthcare in this study. Instantiation is the only type of IT artifact 'tangible' beyond the IT world. DermML is such an instantiation or prototype of a CDSS with the basic features required to use it as a stimulus in this study. It is a real, working system rather than a theoretical or conceptual one.

To classify the stimulus, I adopt a taxonomy according to the decision space that the CDSS support, as below:

1. Focused CDSS that deal with either binary or limited decision options. Melanoma (a lethal skin cancer) detection apps are examples of this type. (Thissen et al. 2017).

2. Speciality specific CDSS that cater to a clinical specialty such as cardiology or dermatology (Tleyjeh et al. 2006). The decision space in this type is broader than focused CDSS.

3. Generic CDSS cater to a wide decision space. Such systems are typically

symptom checkers (Semigran et al. 2015), mostly for patients' use and lack the sensitivity and specificity required for clinicians. It is generally believed that the CDSS is useful to the clinician only when restricted in focus (Sittig et al. 2008).

DermML belongs to the second type according to this taxonomy, as it caters to dermatology. One of the main challenges faced by CDSS in dermatology with a broader scope than melanoma detection is that there are numerous dermatological diseases, some poorly characterized, and many look similar to the untrained eye and to most AI systems. To put this in perspective, a general CDSS in dermatology needs to classify approximately 4000 diseases based on the input received by the CDSS.

DermML belongs to two overlapping subgroups of DSS: i) intelligent decision support systems that use emerging machine learning (ML) and artificial intelligence (AI); and, ii) knowledge management-based DSS. DermML is also an *expert system* because of the following reasons:

1. It relies on a domain-specific ontology.

2. It utilizes an autogenerated, machine-learning-based knowledgebase.

3. The inference is based on the general RDF reasoning.

3.4   Theory

The upfront-theory can have multiple legitimate roles in a qualitative inquiry, such as a source of guidance, a conceptual lens, or a scaffolding (Sarker et al. 2013). I use the TPB as the upfront-theory to develop the interview guide. Figure 3.1 summarizes TPB, and the interview guide is in Appendix B. The questions and

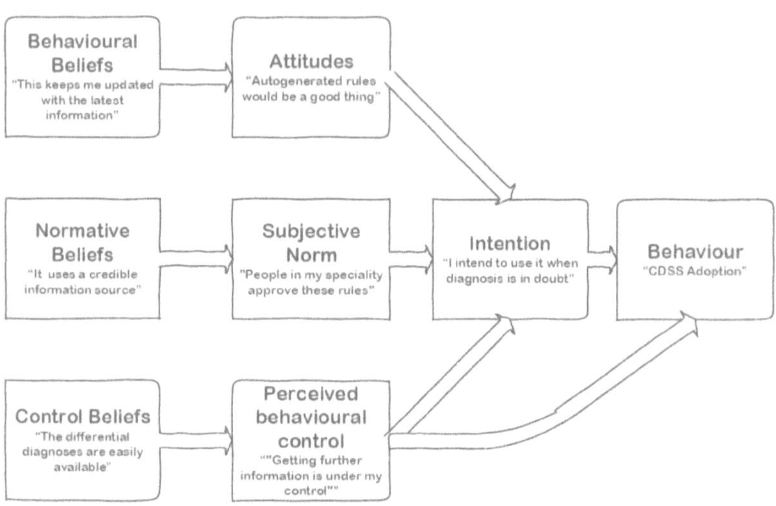

Figure 3.1: The TPB constructs and CDSS

the probes in the interview guide align with the constructs in TPB. For example, I elicit the participants' attitudes and beliefs related to CDSS use.

A stimulus in qualitative research is "something that is shown to the respondents to elicit a response or something that we need to know the respondents' perceptions of" (Törrönen 2002). The upfront theory (TPB) also guided the design of DermML, the stimulus used in this study. The detailed design aspects of the stimulus are described .

3.5 Approach

I adopted the Charmaz (2006)'s constructivist GT for this inquiry and Gioia et al. (2013)'s systematic coding method for analysis. I chose GT as it offers precise techniques and procedures for analyzing data and calls for avoiding apriori theory or pre-existing codebooks. The subjective view of participants and the researchers promoted by Charmaz aligns with my worldview. The next chapter describes the methodology in detail.

At the beginning of the study, I aimed to arrive at a model (a symbolic representation of reality) or a theory (a parsimonious depiction of relationships between constructs within certain boundaries) through analytic generalization. As previously described in the literature review, Gregor (2006) proposed the popular typology of IS theories. Urquhart et al. (2010) suggested that GT can generate any of the five types of theories proposed by Gregor (2006). Inductive approaches are mostly interpretive, while artifacts are real. This study meets at the interface where the reality is co-constructed, where a constructivist GT can make the maximum contribution. I have used the representative quotes of the participants supplemented with my analytic generalizations in presenting the findings.

3.6   Boundaries and constraints

Healthcare traditionally follows a problem-solving tradition that takes the experts' decisions as the gold standard (Patel et al. 2001). However, I specifically go with the decision analytics tradition in the design of DermML and in analyzing the results. In this tradition, experts, as fallible agents, make clinical decisions under uncertainty. This is in line with the shift of AI in medicine towards this tradition and my own experiences with Bayesian approaches (Eapen et al. 2007).

The focus of my work is on the use of CDSS in dermatology. This is becoming popular, especially in rural areas, where dermatology services are commonly managed by healthcare personnel with no specialty training. However, the type of generalizability that I seek is analytical generalizability (Hallberg 2013) within the CDSS in the wider medical domain, though the stimulus itself is restricted to a single medical subspecialty. The notion of reliability, validity and rigour to evaluate this study should be applied with a qualitative and interpretivist lens (Morse et al. 2002).

I investigate HCI at a higher level, based on doctors' general impressions on HCI's importance in CDSS adoption. A formal usability testing of DermML and a detailed exploration of various HCI factors were not within the research scope.

As described before, Hevner et al. (2004) provides a set of guidelines for DSR. DermML as a design artifact complies with these guidelines. DermML is an instantiation that is accessible to doctors through the web. It is based on established constructs, models, and methods (see Appendix A). It solves the problem of cognitive overload faced by doctors in certain decision-making situations.

DermML is projectible to the desired problem space (Baskerville and Pries-Heje 2019) of CDSS. Specifically, the weighting of clinical rules and ordering of differentials tries to move the machine learning world closer to the CDSS world according to the insights given by Goodman (1978). The clinical validity of this method remains to be established. This **book** communicates the findings to the business-oriented audience in the form of a theory and the technology-oriented audience in prescriptive design statements (Baskerville and Pries-Heje 2019).

I extend the guidelines for applying GT in a pluralistic context. I posit that my approach helps researchers in avoiding certain potential pitfalls while applying a method like GT with a well-established set of guidelines, principles

and techniques in a multi-paradigmatic setting (Gregory 2011). In this study, I
use a real (functional) stimulus that serves two purposes. It elicits a response
from the participants that help me to move towards a substantive theory in
CDSS and to project the findings of my study to the specific problem space —
diagnostic decision making.

In summary, the literature review shows that a substantive theory for CDSS
adoption that addresses the many clinical decision-making challenges is needed.
Knowledge augmentation systems may require a different evaluation method that
considers design.

Chapter 4

Methodology

4.1 Introduction

Context is deeply ingrained in CDSS use, and a qualitative exploration may be better suited to study the phenomenon of CDSS use. I adopted the GT method for the qualitative exploration of CDSS. The choice of dermatology as the clinical specialty for the study is based on my domain expertise. First, I briefly introduce GT focusing on the Charmaz (2006)'s constructivist GT and reasons for adopting this genre of GT, followed by the description of data sources, stimulus, sampling and interview methods. The study protocol is summarized in Figure 4.1.

4.2 Grounded theory

GT inductively develops insights from data without relying on an upfront theory or a set of hypotheses (Glaser and Strauss 1967). The research questions tend to be broad in GT, and a wide range of data is triangulated, iteratively conceptualized and scaled up to reach a substantive theory. GT emphasizes theory building as opposed to theory testing in most quantitative methods (Urquhart et al. 2010). GT can also be used as a coding technique within other qualitative research methods (Strauss and Corbin 1990).

GT is suitable for studying repeated processes and for investigating phenomena that do not have a well-accepted explanation or a theory. It is also useful in situations where quantitative methods have been less successful (Kaplan and Shaw 2004). New researchers usually find it easy to analyze the data early in the research process, since this provides evidence to formulate an early hypothesis. It also encourages a constant interplay between data collection and analysis (Myers 2009). GT has been used to assess the patient-perceived usefulness of online electronic medical records (Winkelman et al. 2005). It has also been used to study how consumers search for health information on the internet (Eysenbach and Köhler 2002), and for theory generation related to HIS security threat lifecycles (Fernando and Dawson 2009).

4.2.1 Constructivist GT

GT practitioners have a pragmatic worldview and believe in practical theory (Cronen 2001). The founders of GT, Glaser and Strauss, later diverged in their approach, with the former emphasizing induction (Glaser 2002) and the latter emphasizing a systematic approach (Strauss and Corbin 1990). Glaser's GT describes two stages of coding: substantive coding and theoretical coding. Substantive coding can be subdivided into open and selective (Walker and Myrick 2006) types. Glaser's GT encourages the triangulation of various data sources. Strauss and Corbin recommended a strict coding structure elaborating on how to code and structure data. The seminal article by Strauss and Corbin

(Strauss and Corbin 1990) describes four stages of coding: open, axial, and selective coding followed by a 'coding for process.' There is no explicit mention of theoretical coding.

Charmaz (2006)'s constructivist approach treats GT research as a construction, acknowledging the involvement of the researcher in the construction and interpretation of data. The researcher is not a passive onlooker anymore and is very much part of the research, bringing in his/her/their theoretical sensitivity (Glaser 1978). Charmaz's constructivist GT methodology belongs to the second-generation qualitative research approach that has not yet been firmly established in IS discipline (Wiesche et al. 2017).

GT is not without drawbacks. Coding in GT may be overwhelming, and scaling up of categories from open coding may be difficult. This leads to the generation of low-level theories. To mitigate this problem, I adopted Gioia et al. (2013)'s systematic coding method to generate 1st order concepts that lead to 2nd order themes and, finally, aggregate categories. More information about Gioia's (2013) analysis method is described further below.

The basic tenet of GT is abductive reasoning and qualitative induction (Reichertz 2007). Popper et.al (Popper 1979) and Wirth et.al (Wirth 1999) put forth the concept of abductive reasoning, a process of backward reasoning from the observation of a surprising fact that fails an expectation. Zmud et.al (Zmud 1998) recommends the following steps for inductive theory building:

1. The description of the phenomenon

2. The construct creation, development and explication

3. Identification of key relationships

4. Development, justification and articulation of these relationships.

I adopted Charmaz's constructivist GT (Charmaz 2006). My choice of constructivist GT is based on the four dimensions proposed by Sarker et al. (2018): i) nature of data, ii) nature of theory, iii) nature of analysis and iv) nature of claims that I wish to make. My data include interview transcripts, demographics and clickstream data triangulated holistically by systematic analysis to arrive at a substantive theory for CDSS. The objective of making theoretical and practical contributions supports a constructivist approach. Charmaz (2006)'s constructivist GT that I adopted has a subjective view of data and an interpretivist view of theory (Mills et al. 2006). Charmaz (2006) stresses the role of the researcher in the shared construction of reality. I belong to the clinical community to which all my participants belong, and as such, my theoretical sensitivity influences my analysis and claims.

4.3   The study protocol

I invited dermatologists to view a demonstration video of DermML and use the system. Interviews were scheduled with consented participants after a week of DermML use. I demonstrated DermML use before each interview, and the semi-structured interviews were audio-recorded. I collected participant details after the interview and analyzed it with the audio transcripts and clickstream data. The study protocol is detailed below and summarized in Figure 4.1.

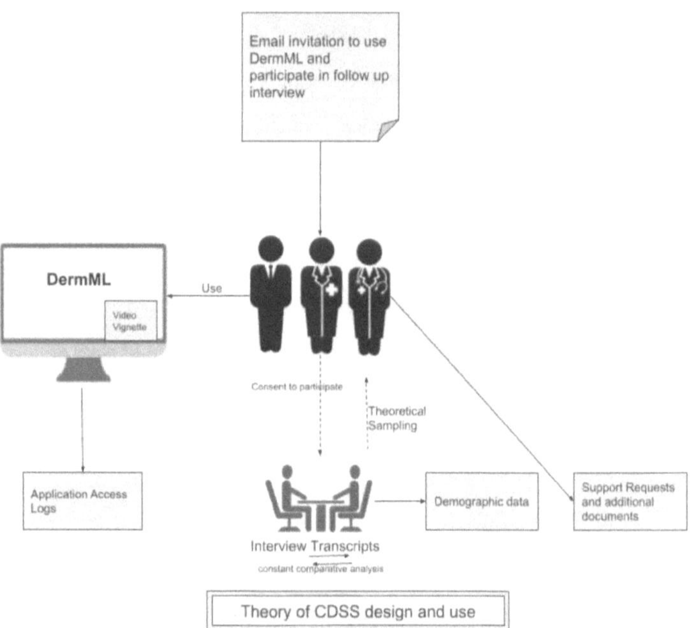

Figure 4.1: The study protocol

4.3.1 Population sample and recruitment

The study was approved by the McMaster Ethics Review Board (MREB). Email messages approved by the ethics board were sent to a few dermatologists, giving them a brief background of the study and inviting them to use DermML for hypothetical treatment scenarios and for learning purpose. The subsequent theoretical sampling strategy is described later in this section. DermML displayed a prominent banner stating that it is not intended to be used for the diagnosis or treatment of patients.

4.3.2 The stimulus

Using an IT artifact as a stimulus, without the backing of a theory in a qualitative enquiry, is unlikely to contribute "anything more than a chaotic bundle of statements, impossible to decipher or evaluate or apply to any meaningful purpose" (Harrington 2005). I use the same upfront theory — TPB — as a guide for data collection in the form of an interview guide and DREPT to guide artifact design. TPB was chosen as it is less abstract compared to its predecessors, such as TRA, while being less complex than its successors such as UTAUT. The up-front theory was not used during the data analysis, for which I followed the Gioia et al. (2013) method. I believe that using the same kernel theory for qualitative study and guiding the artifact design is central to this study.

4.3.3 The vignette

It is challenging for users to start using DermML as they are not familiar with the interface. To overcome this barrier, I presented a video vignette in DermML that demonstrated the purpose of various interface components and explained their function. The vignette depicted carefully designed realistic scenarios on DermML use. The vignette also described DermML design, how it extracted knowledge from published literature and how domain experts could collectively vote and refine the rules. After viewing this information, users could observe an ideal scenario where DermML might be useful as a diagnostic aid. The participants always had full access to DermML and the vignettes.

4.3.4 Interviews

The invited users who were willing to participate in the interview were contacted after one week of DermML use, and a suitable time was set up for an interview. In the initial phase of the study, face-to-face interviews were conducted with willing participants. The audio recording was made during the face-to-face interview after taking consent. Later, interviews were moved to the Zoom platform as face-to-face interviews were discontinued during the COVID-19 pandemic. Approval was sought and obtained from MREB for online interviews using the Zoom platform, and the consent process was changed to suit the online interviews. All recordings were made using the built-in recording function of Zoom. The interviewee's video was disabled by default during the interviews, and only the audio was recorded for transcription and analysis. GT does not prescribe an ideal sample size. Hence, I started by interviewing a few clinicians and continued theoretical sampling (described later in this section) until theoretical saturation was reached after 14 interviews.

Semi-structured interviews were conducted with the help of an interview guide . The interview guide was used when the participants needed a trigger to continue the conversation. The participants were encouraged to discuss 'anything that came to mind regarding CDSS.' I demonstrated the use of DermML to each participant before the interview, as detailed in the next section. On average, interviews lasted 40 minutes, including 5-8 minutes of the demonstration of DermML. (Details in the results section)

The analytical memo was maintained after each interview by summarizing the findings thus far. The analysis and constant comparison as prescribed by GT was followed, and purposeful sampling was performed. Some structured data was collected during the interview. The transcription was performed manually by me. All personally identifiable information was removed from the transcripts.

4.3.5 Demonstration

As mentioned before, I demonstrated the use of DermML before each interview. The demonstration audio was recorded and transcribed along with the interview to test if any minor changes in demonstration content had any impact on the participant response. The demonstration included all the steps and content in the video vignette.

I adopted the input-process-output-engage (IPOE) model (Khairat et al. 2018) for user engagement during the demonstration. I emphasized that the goal of DermML is to supplement the doctors' decision-making ability and not to replace the doctor. The input was specified as the positive clinical history and findings in the free text form and the lesional images. The multi-modal machine learning process and the algorithm for generating differential diagnoses were explained at a high-level to every participant. The expected output, such as the list of differentials, relevant evidence and the weightage score, was mentioned. Some of the specific output characteristics related to the order in which the differential diagnosis list is presented to the user (rare conditions first) were emphasized. I clearly stated during the demonstration that DermML is a passive CDSS that provides information to the doctor, but does not record, mandate, acknowledge or request action by the doctor (Musen et al. 2006).

4.3.6 Reflexive memo taking

I used reflexive memos following a simple framework focusing on three main aspects (Srivastava and Hopwood 2009):

1. What are the data telling me?

2. What is it that I want to know?

3. What is the dialectical relationship between what the data are telling me and what I want to know?

The reflexive memo template.

4.3.7 Usage data and other data sources

The interaction of users with the application is an important source of data. Clickstream analytics is the analysis of the usage pattern --- which pages are commonly accessed, in what order and the duration of interaction — mostly for usability and e-commerce analytics. Most clickstream reporting tools record some personally identifiable information about the user, such as the IP address. However, this poses a privacy problem, and hence I developed a custom backend to record these details, avoiding personal information. A sample of anonymized clickstream data

4.3.8   Interview guide and data collection instrument

The interview guide is based on the constructs from sensitizing theory. The constructs in TPB are perceived behavioural control (PBC), attitude and subjective norm. PBC was assessed by probes such as 'value of CDSS system' and the 'feelings related to using the system.' The attitude was assessed by asking about the 'experiences with CDSS,' and the subjective norm is reflected in the attitude towards the voting module and social media dissemination of *system use* related messages.

Demographic data such as the gender, age and type of practice were collected from the interviewee in a separate form . This form also includes meta-data about the interview, such as time/date of interview, its duration and the name of the audio file.

4.3.9   Data collection

To summarize, I collected the following data as described above:

1. Anonymous clickstream data from the DermML application, such as commonly accessed pages, number of visits and the duration of the visit by each user.

2. Any support request sent from DermML. The email address and any

personally identifiable information were removed from the message before analysis.

3. The demographic details of the participants were recorded prior to the interview.

4. Interviews of participants selected following the theoretical sampling technique.

4.3.10 Coding and analysis

All interview data were recorded using a sound recorder during the initial face-to-face interviews and by the Zoom platform's built-in recording function later. I transcribed all interviews personally, which helped me to reflexively analyze the data after each interview. Coding identifies categories in the collected data. Coding was performed using the Dedoose online software (Lieber and Weisner 2013) following the Gioia et al. (2013) method.

The Gioia method

I use the data structures introduced by (Gioia et al. 2013) to bring "qualitative rigour" to my research but with certain caveats to modulate it for the phenomenological needs (Bansal and Corley 2011) of my investigation which in the true tradition of IS inquiries have an artifact at the centre. Gioia et al. (2013) proposes a three-staged process-oriented analysis in which the core concepts are identified in the first stage, subsequently distilling them into broad 2nd-order themes. Finally, aggregate dimensions are identified which are scaled to a model or a theory. Gioia et al. (2013)'s approach is a mix of Strauss and

Corbin (1990)'s open coding (1st order concepts), axial coding (2nd order themes) and Glaser and Strauss (1970)'s theoretical coding (aggregate dimensions).

The ground assumptions of Gioia et al. (2013)'s approach is based on an organizational socially constructed world with knowledgeable agents. I take a different view that part of the IS world is real, and some of the findings can be formulated in practical terms, in addition to the theoretically relevant terms. Methodologically, my different epistemological view leads to the following differences:

1. The term *data structure* used by Gioia et al. (2013) may confuse IS researchers with a computer science background as it represents a different and well-established concept in computer science. I use the term *analytics flow chart* to represent the same concept. I use the term *aggregate category (or just category)* instead of *aggregate dimensions*, as I have used *properties* and *dimensions* in a more traditional way to represent attributes common to all concepts in a category and the observed variations of a property, respectively (Moghaddam 2006; Urquhart et al. 2010).

2. I organize 2nd order themes and the aggregate categories into three types: (a) Contextual, (b) prescriptive and (c) combination. These subdivisions are colour coded in the analysis flowchart (see Figure 5.7), and the significance is explained in detail later.

3. I propose two models: one to describe user adoption in the substantive domain of CDSS and the second, prescriptive propositions to guide design. The first one belongs to Gregor (2006)'s type 4 theory — for explaining and

predicting CDSS adoption, and the second one belongs to the type 5 theory — for design and action. In short, in addition to a set of propositions as in the original Gioia et al. (2013)'s approach, I propose prescriptive guidelines for CDSS design.

The constructivist GT relies on the researcher's interpretation of the data. Hence, I avoided word-to-word coding at the open coding phase and emphasized emerging 1st order concepts. I made an effort to use the participants' special terms (in vivo codes) whenever possible (Charmaz 2006) during this phase. I tried to minimize the impact of the sensitizing theory in the initial coding phase. This was to ensure that I remained open to all theoretical possibilities. However, in my constructivist approach, my awareness of the sensitizing theories and theoretical sensitivity may have influenced data collection, analysis and theory building.

I refined the 1st order concepts to 2nd order themes and finally to aggregate categories at the focused coding stage. I used the constant comparative method at all stages. Next, I used axial coding to identify the attributes and dimensions of the categories that I identified. I followed Charmaz (2006)'s style of axial coding and not the Strauss and Corbin (1990)'s framework that identifies predefined links between categories during axial coding. Memo-writing is an integral part of GT. I maintained reflexive memos during data collection and coding to aid theory development. The template for memos . Finally, I scaled my codes at the stage of theoretical coding, identifying the boundaries and relationships, and arrived at a substantive theory. As mentioned before, I also proposed a set of prescriptive guidelines (type 5 theory) for CDSS designers.

In GT, the researcher is expected to reveal the researcher's background to allow others to assess the validity of findings and accounts. I had a transformational view that CDSS fail because they are predominantly tested in places where CDSS are less required. I believe in the emancipation of resource-deprived areas of the world through technological innovations in healthcare. My findings point in this direction (CDSS users and its use are heterogenous) but did not substantiate my views.

4.3.11 Pilot

During the pilot phase, I interviewed one dermatologist using the same interview guide. Shortcomings identified during this stage were rectified before the main study. I did not conduct a separate pilot after changing interviews from face-to-face to virtual on Zoom.

4.4 Assessing contributions

It is customary in qualitative research to recommend a method to assess the research contributions. I explored the socially constructed realities of CDSS use, shared by a group of clinicians to which I also belong, with an interpretivist ontology and a subjective epistemology. My analysis was not particularly data-centric, with my interpretations playing a major role in constructing the shared reality. My aim was to generate a theory that defines variables, relationships between variables, the justification for the relationships and the boundary conditions. If the boundaries and the justifications for the relationships between variables were not clear, I hoped to build a model or a rich description of the phenomena (Wiesche et al. 2017). However, the relationships and the boundaries for the core categories that emerged were clear and generalizable to the substantive area of CDSS. Hence, as described later in this book, my contribution is a substantive theory for explaining and predicting the user adoption of CDSS.

Qualitative research's evaluation criteria depend on the genre because of the differences regarding how each genre approaches data, theory, analysis, and claims (Sarker et al. 2018). As previously mentioned, I have adopted the constructivist GT genre with an emphasis on inductive theory building (Charmaz 2006). Sarker et al. (2018) recommend the following criteria for evaluating research that belongs to the GT genre:

- Reliability of data

- Theoretical sampling procedures

- Systematic coding

- Constant comparative analysis

- Theoretical density

Next, I describe how this study addresses these aspects along with projectability to the problem space as an added criteria. I explained the notion of projectability in my foregoing literature review (Baskerville and Pries-Heje 2019).

4.4.1   Reliability of data

In qualitative research, the notion of reliability aligns with consistency and not exact replicability as in quantitative methods (Leung 2015). I interviewed all clients and transcribed the data myself to maintain consistency. My familiarity with the domain of inquiry, as described previously, helped the data collection process. I kept reflective memos during the study to assist with scaling the findings into a coherent theory.

4.4.2 Theoretical sampling procedures

DermML belongs to the category of diagnostic CDSS for dermatology. Hence, I started my research by interviewing a sample of dermatologists, iteratively analyzing the data I collected, and maintaining reflexive memos after each interview. However, I soon realized that dermatology wait times are long, and access to dermatology services is scarce. In fact, I found that most dermatology cases are managed initially by the family physicians. The initial participants suggested that a dermatology CDSS is far more likely to be used by a family physician. I noticed this dimension of 'system user' emerging during my initial interviews and started including family physicians in my sample who routinely see dermatology cases. The importance of machine learning-based CDSS in other visually-oriented specialties such as radiology and ophthalmology emerged later, and I included a couple of doctors from these specialties to get their views. Overall, I interviewed 14 participants and recorded 780 interactions with the CDSS application.

4.4.3 Systematic coding

I used Gioia et al. (2013)'s systematic coding procedure to progress from the 1st order concepts in the open coding stage to 2nd order themes, finally leading to aggregate categories. I used some known construct names in IS as the category names. The use of the same terms does not imply that both are the same, but the similarity helps theoretical integration.

4.4.4   Constant comparative analysis

Iterations are reflexive processes in GT that help to gain insight from the already collected data and to guide further theoretical sampling and to move towards theoretical saturation. I used constant comparison at every stage. This helped in achieving the state of theoretical saturation through theoretical sampling, as mentioned before. I kept an open mind during the final analysis following Gioia et al. (2013)'s method. However, many of the categories I saw emerging during the study proved to be right. I transcribed all the interviews personally, and in my opinion, the process of transcription leads to subconscious iterative comparison and reflexivity. Transcription quality is shown to be a vital aspect of rigour in qualitative research (Poland 1995).

4.4.5   Theoretical density

I propose two theories; one for explaining and predicting user adoption of CDSS systems and a second for prescriptive design theory. The scope, the level of conceptualization and the theoretical integration of both theories are detailed in the discussion section.

Gioia et al. (2013)'s method was beneficial for inductive theory building. My theoretical sensitivity in this domain also helped in scaling participant statements into a coherent and parsimonious theory.

4.4.6 Projectability in the problem space

The notion of generalizability has limitations in the realm of DSR. (Baskerville and Pries-Heje 2019) proposes projectability to assess design contributions. The prescriptive design rules generated in this study are projectable into the problem space of multi-modal machine learning in visually intensive medical specialties.

Next, I will describe the data and the results obtained.

Chapter 5

Data and Results

"Not everything that can be counted counts and not everything that counts can be counted." - Albert Einstein

5.1   Introduction

As described previously in the methodology section, the participants for the semi-structured interview were doctors with experience in dermatology and other visually oriented specialties. This included dermatologists, family physicians, ophthalmologists and radiologists. The participants were invited by email that included a link to DermML accessible through the internet. The participants were encouraged to view and explore DermML before the interview.

The study used four distinct data sources:

1. Before the interview, a few optional demographic and practice related data were collected using a questionnaire.

2. All user interactions with DermML were recorded anonymously using a

Figure 5.1: Demographic distribution of interview participants.

custom script.

3. Any communication that participants sent after the interview, directly related to DermML.

4. The semi-structured interviews with the participants, face-to-face and virtual, audio-recorded and transcribed.

5.2 Demographic data

The total number of interview participants was 14. The questionnaire data showed an even gender mix with eight females (57%) and six males (43%). Half of the participants revealed their age to be above 40. (See Figure 5.1).

Eight participants (57%) had previous experience practicing in a low resource setting compared to Canada, and half of them had previous experience with a CDSS system. The participants were encouraged to view the demo video available on the DermML website to familiarize themselves with the many features of DermML before the interview. However, only six participants (43%) mentioned that they did watch the video. These results are summarized in

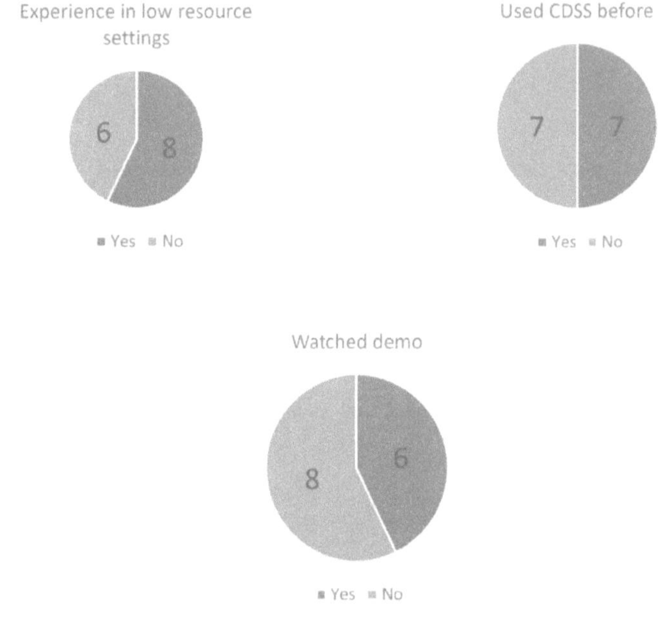

Figure 5.2: Summary of participant experience

Figure 5.2.

As the stimulus used is a dermatology CDSS, I started my study by sampling dermatologists. Later, aligning with the concept of theoretical sampling (sampling based on where the theory takes me), I also included family physicians with experience seeing dermatology patients. In many parts of the world, including Canada, dermatology specialists are in short supply, and common skin conditions are first managed by family physicians (Stephenson et al. 1997). This emerged during the interviews, along with the need to include specialists from other visually oriented medical specialties such as radiology and

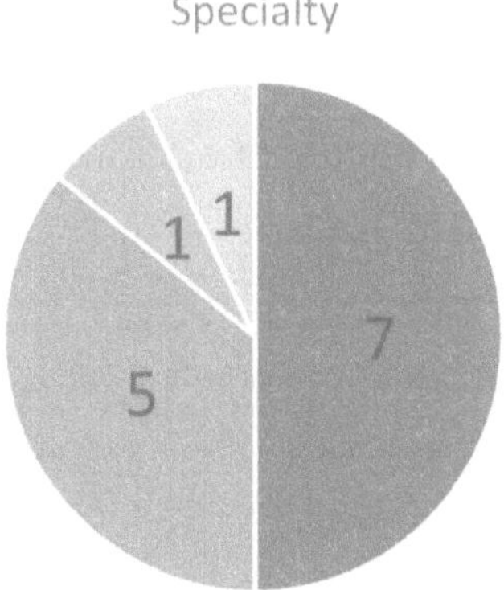

Figure 5.3: Area of expertise of participants

ophthalmology. Hence, the final sample included these specialties also, and the distribution is shown in Figure 5.3

5.3   Interviews

I completed 14 interviews, each lasting approximately 45 minutes. Though the results below are presented in a logical sequence from 1st order concepts to 2nd order themes and then to aggregate categories for sensemaking, the analysis and reflective memo writing happened throughout the study in an iterative manner. Some of the aggregate categories, such as *system user* and its various dimension discussed later, emerged early in the study. This led to a *theoretical sampling* process to include some of the potential *dimensions* that I would have otherwise missed. I personally interviewed all participants and transcribed the audio.

5.4   Analysis

After reaching the stage of theoretical saturation, a second comprehensive analysis of transcript data was performed in its totality using Dedoose (Lieber and Weisner 2013). Three hundred thirty-three excerpts were highlighted from 14 interviews (the total word count of all transcripts was approximately 75000). Close to 80 1st order concepts were identified while coding these excerpts. I attempted to represent these 1st order concepts with the participants' own words as much as possible (See analysis flowchart in Figure 5.7). These 1st order concepts were iteratively conceptualized to twenty-seven 2nd order themes (hereafter themes) and five aggregate categories. The 2nd order themes are summarized in Table 5.2.

Five aggregate categories that emerged from the systematic analysis were *system user, use context, information quality, knowledge presentation and technology efficacy.* The system user and use context categories are descriptive or explanatory categories with properties and dimensions that do not have any direct impact on system design. Hence, both system user and use context are grouped as *contextual types.* The information quality and knowledge presentation have both explanatory as well as design properties, and as such, both are grouped as *combination types.* For example, the *knowledge presentation* may influence PBC and provide prescriptive guidelines for presenting information in a CDSS. The technology efficacy category is a *prescriptive type* category with direct implications on design. The aggregate categories and their types are summarized in Table 5.1.

Table 5.1: Categories that emerged from interview analysis and their types

Type	Aggregate category
Contextual	System user
	Use context
Combination	Information quality
	Knowledge presentation
Prescriptive	Technology efficacy

I followed the Gioia et al. (2013)'s systematic method for the analysis of interview transcripts. Next, I describe how these categories emerged with respect to their 2nd order themes, and the 1st order concepts directly from the interviews. Heeding the Urquhart et al. (2010) call to "putting the theory back into grounded theory" I then tease out *properties* of these categories using the process of iterative conceptualization (Urquhart et al. 2010). Finally, in the discussion section, I explore the relationships between the categories and their properties and scale into type 4 substantive theory for CDSS adoption and a type 5 prescriptive theory for CDSS design.

5.4.1 System user

I have used system user as a category to represent stakeholders who use the CDSS for diagnostic decision-making. As explained before, I have used the term "doctor" to represent this group in the earlier discussion, but decided to use an IS centric *system user* category for coding to help theoretical integration later. Some of the themes that support this category are listed in Table 5.2 and explained subsequently. I found it interesting that some participants used the pronoun 'they' instead of 'us or we' though participants themselves belonged to the groups they describe.

Table 5.2: Themes associated with the system user category

Category	Themes
System user	Cognitive load
	Learning new information
	Practitioner
	Change Management
	Threat to professional autonomy
	Socio-cultural

Cognitive load

"I would quickly want to diagnose them and go to the next patient, then the next one. Keep Moving on..."

Cognitive load is a common theme that emerged with 7 of the 14 participants mentioning the importance of CDSS in reducing the cognitive load involved in decision making. Interestingly males mentioned this more than females. The cognitive load is a known patient safety factor in the context of intensive care units (ICU) (Faiola et al. 2015). Many of the participants feel that increasing cognitive load is a growing concern because of the increasing information overload from various data sources, a concern that I personally share. I was not surprised to find that 66% of those who mentioned this had experience practicing in a low resource setting.

"Sometimes, the clinician may forget some of that knowledge gained during education and experience, will definitely help him, remind that okay, this is the condition."

It is hard to retain all the information needed for patient care. In my opinion, an increased cognitive load may be due to a complex presentation or due to an increase in the number of patients within a limited period (Harry et al. 2018). CDSS can potentially reduce the cognitive load in both situations.

"searching just can give to me immediately an idea about what I'm dealing with."

The capability to search efficiently is vital in reducing cognitive load. Interestingly, many respondents mentioned the popular search engine, Google, in different contexts. Intelligently executed searches on the internet can deliver important and timely information for decision support. However, it is difficult for doctors to design a good search strategy while engaged in patient management. Besides, most of the results returned are likely to be generic, lacking credibility and depth required for clinical decisions.

Learning new information

"This is like an ongoing CME [Continuing medical education]"

Most participants (11 of 14) emphasized the role of CDSS in learning new information, outside a decision-making context. The utility for residents or trainees in healthcare for learning new information was an important theme that emerged. This was shared more by dermatologists (47%) than family physicians (30%). Surprisingly, the role of internalization also surfaced. Knowledge gained during a *"leisure learning"* process may be useful in a decision making context later.

"And usually we put these things like erythema and itchy plaques, we'll put these on Google, but it could not come up with the real differential itself. it's an ocean [search engine], and it's not very specific."

CDSS facilitate learning and internalization of knowledge, unlike search engines such as Google that provide a lot of non-specific information. Learning is facilitated by a more focused source of information (Vandenbosch and Higgins 1996).

"Every day, they [patients] challenge you. Yes. Even in the cosmetic field. There is a lot of material online, and they [patients] just come in and challenge you and say that's not right."

One of the factors that motivate doctors to update their knowledge is the pressure from patients. The easy availability of health information online has empowered the patients. However, many doctors feel that patients are generally incapable of separating the grain from the chaff. CDSS mediates the knowledge update for doctors.

Practitioner

"Especially our senior doctors, they have a specific mindset and from their experience, and they don't rely a lot on clinical decision support."

It is widely believed that older doctors are less likely to adopt HIS (Abdekhoda et al. 2015; Or et al. 2014). Hence, it was no surprise that seniority emerged as an important theme. However, participants did not seem to distinguish professional seniority from age. Hence I conceptualize age, professional seniority and type of practice under this theme.

"I found less resistance from the older generation was gonna surprise for me, as compared to like you know, the younger generation."

I was surprised to find the above contrasting observation from one of the participants. This is probably related to the confusion between professional and chronological seniority.

"For a young resident or trainee, for example, going to university where they're going to train the residents with it ..... it gives you diverse information without actually having to go for it, sitting at your fingertip, so you can see the correlations, as well as rare feature presentations, and treatments that work without having to search."

It is a formidable challenge for residents undergoing training in a particular specialty to filter the deluge of new and often contradictory information frequently added as *evidence*. In my opinion, young doctors, who have not yet developed the tacit knowledge required for diagnosis, are more likely to adopt CDSS.

"I'm thinking about the older professors, it's good for the younger generation, but for the older professors whom they depend upon their experience. [it may not be suitable]"

Older doctors may view CDSS as something that can potentially "flaw the logical process of their decision making" (Davidson et al. 2013), probably a form of cognitive dissonance that may not be in the best interest of patients.

Change management

Four participants mentioned the importance of change management in the adoption of CDSS.

"It's a kind of change management. Usually, people are resistant to change."

This is a common theme that emerged in many previous studies (Davidson et al. 2013; de Grood et al. 2016; Greenes et al. 2018; Yusif et al. 2019). CDSS implementations lead to a change in the clinical workflow. This has been identified as one of the barriers for implementation failures(Yusif et al. 2019).

"They just like to do what they're used to doing. And give them results. You have to compel them to see what are you bringing more to the table?"

From a doctor's perspective, this may be due to the desire to make their own decision without the "bias" introduced by CDSS (Davidson et al. 2013; de Grood et al. 2016). The importance of educating doctors about the need for change in better patient management cannot be understated. It is widely recognized that *clinician champions* play an important role in facilitating change (Chrusciel 2008; Greenes et al. 2018; Mount and Anderson 2015), a theme reiterated by one of the participants.

"... you need to have those champions prior [to implementation]."

Canada Health Infoway has a change-management framework to understand the process of facilitating change associated with HIS and a toolkit to help implementers with the process (Catz and Bayne 2003).

As one participant nicely summed up: *"Physicians are very, very reluctant to use new technology,"* and in my opinion, that has to *change*.

The threat to professional autonomy

"They like their autonomy; they like to feel that they're in control of the practice."

The perceived threat to professional autonomy (PTPA) is a commonly studied construct in the context of HIS adoption (Esmaeilzadeh et al. 2015; Sambasivan et al. 2012; Walter and Lopez 2008). From an IS perspective, PTPA has known negative effects on PU and thereby intention to use (Sambasivan et al. 2012; Walter and Lopez 2008), and this effect is larger for CDSS than EMR in a healthcare context (Walter and Lopez 2008).

"Because most of them think that okay, they have a system now, then, later on, they wouldn't need us even, which is a valid point, but still, patients require human contact. It's not just a PC [that patients require]."

Some of the factors that can reduce this threat are perceived interactivity and the attitude to knowledge sharing (Esmaeilzadeh et al. 2015).

"They like their autonomy; they like to feel that they're in control of the practice."

"We're not computer science [professionals]. We'll are not IT people, who don't know what your [IT professional's] stuff is"

One known way for reducing PTPA and thereby increase adoption is to involve doctors in the planning, design and implementation, a theme that I will discuss later. Performance expectancy — the belief that CDSS will improve job performance — increases intention to use (Esmaeilzadeh et al. 2015; Sambasivan et al. 2012).

Socio-cultural

"I'm not gonna take someone else's opinion unless I know him."

The influence of socio-cultural factors in technology adoption is undisputed, and CDSS is no exception. However, in this study, I have used this category in a restricted way to represent the impact of social media on CDSS adoption. During my years of clinical practice, I used to maintain a Facebook group for dermatologists, and over the years, I have seen and moderated many discussions related to CDSS adoption. This motivated me to study this relatively less explored dimension.

"We're now in the COVID-19 era. Social media is very important for us as clinicians interact with [it]. For example, we have a WhatsApp group. And we use this and Google messenger to help each other. [We] share articles on COVID 19, lots of information, updates on clinical conditions, updates on guidelines, change in protocols and treatment. Social media for clinicians provide very real-time [platform for] information transfer and knowledge transfer on the spot."

As expected, the participants highlighted the importance of social media in their decisions. Nevertheless, some were skeptical, and the need for restricted doctor-only groups or an internal social media platform emerged as a theme. Bittner et al. (2019) has described the advantages of such closed social media groups and proposed an informed consent template for educating and protecting patients.

"An internal platform like emails from your leadership would be really helpful. I don't see the common social media to impact health care providers' decision to [use CDSS], you know, I don't see that as a major impact."

"Social media is open to everyone. We need specific tunnelling to ensure it's all dermatologists there."

Patient privacy — a well-known factor in social media interactions of healthcare professionals (Melnik 2013; Ventola 2014) — was identified as an important barrier.

"Patient information sometimes inadvertently, post a picture and sometimes an identifying picture comes in, which can put physicians in big trouble and there might be the lawsuit coming your way."

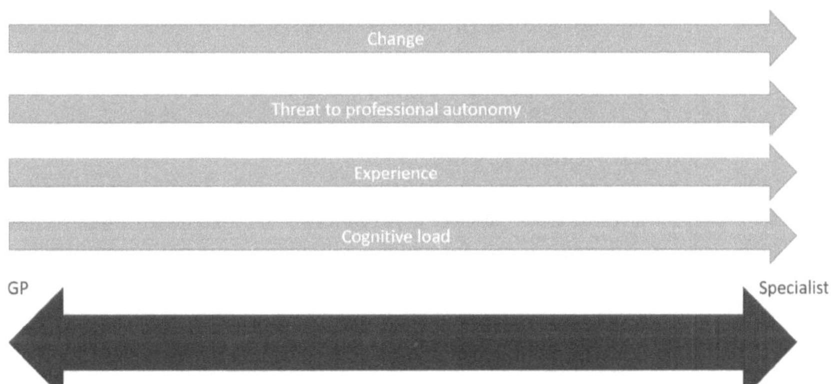

Figure 5.4: The dimensions of 'system user' category

Properties and dimensions

Some of the properties associated with the system user category are academics, trainees, residents and senior practitioners. The healthcare sector is diverse, and these terms have different meanings in different regions. Hence, I propose GP (general practitioner) vs specialist as two dimensions of *system user*.

I use GP — a universally recognizable acronym — to represent a doctor whose practice is not oriented to a specific medical specialty but covers a variety of medical problems for all ages, but of relatively low complexity. I have used *family physician* --- a title commonly used in North America — as a synonym for GP and used both interchangeably in this **book**. The bidirectional arrow indicates the continuum of the doctors' experience. Though the relationship of many of the themes with this dimension is debatable, I use this continuum to broadly represent their direction of variation (see Figure 5.4) and sum up with another participant comment.

"If this system is given to a family physician or a general practitioner, it will come in very handy."

5.4.2 Use Context

Use context is the category that represents the conditions under which the doctors use the CDSS. The use context is the chief factor that sets apart CDSS from other IS. Figure 1.1 shows the complexity of the context in which CDSS systems are typically used. Understanding and characterizing the peculiarities of the context of use is vital in the successful adoption of CDSS systems. The themes that support this category are listed in Table 5.3.

Table 5.3: Themes supporting the use context category

Category	Themes
Use context	Healthcare setting
	Confirm decision
	Reduce errors
	Efficiency
	System Properties

Healthcare setting

CDSS may be used in a variety of settings such as hospitals, clinics, emergency rooms, multispecialty groups, etc. and the use varies according to the context (Heavin 2017). With the growing cost of healthcare delivery in most countries and the cost associated with medical training, some medical specialists such as dermatologists are becoming increasingly rare. This leads to increasing wait times for specialty consultations, and many such cases are managed by general practitioners (Ehrlich et al. 2017). This concern was echoed by many participants, and the role of CDSS in such situations was emphasized.

"We have a large influx of patients, especially in the outpatient department. So I need to make these decisions quickly. You know, so this is going to assist me in all the soft areas. And then sometimes you don't have enough dermatologists, so we rely a lot on the general practitioners, you know to make use of this system, so it should be very handy for the general practitioners to be able to make these diagnoses on time. And probably more accurately."

Participants made a clear distinction between the potential use of a CDSS in an academic setting with an emphasis on information search. Some also alluded to the possibility of offloading some of the teaching efforts to a CDSS, to reduce their workload. Ommaya et al. (2018) recommends appropriately designed CDSS for reducing physician burnout.

"In an academic setting, it would be a very good educational tool for residents. If the attending physician is a consultant physician, he may not be directly using it, but as a resident, he goes inside the room before talking to the physician to whom he is presenting the case, they would put in these and try to look at everything [differentials] and then come up with something and when they discuss it, [in a] much more useful discussion, and in the process, the consultant [is] indirectly learning. Yes, it is much more interactive, and it's not like babysitting the residents or medical students. They are also stimulating you, and you become more interested. I think it becomes a much better relationship and a much more useful interaction."

Treatment information is available on DermML, and I explored if there is any preference for the type of support that doctors prefer. The consensus was that diagnosis is the primary step that needs cognitive support. It is easy to extend that information to find the right treatment.

"It's for consultants with experience; they would use it for therapeutic purposes. Once you get the diagnosis, right, you can find the treatment."

Confirm decisions

"I second guess myself a lot."

Doctors often use CDSS to confirm their original diagnosis. This theme (unlike the next related theme; reducing errors) was mentioned by dermatologists (specialists) more than family physicians, an important aspect in our later discussion.

As previously mentioned, bad decisions in healthcare can potentially lead to loss of human life (Patel et al. 2002). Experienced doctors rely on initial hunches that they develop during the encounter with a patient(Davidson et al. 2013). But such initial hunches are liable to availability bias that occurs when physicians diagnose cases that have similarities with recently encountered cases (Mamede et al. 2010). The effect of such availability bias was a common theme that emerged in this category.

"I sometimes think we sort of fixate on certain things. I do remember when I was a resident, some clinical diagnoses, and just come up to mind, for example, psoriasis first, and then I have to check myself. Are you sure? So as a savvy clinician, you want to tell them that you are critical. .... the system brings back the widespread differential diagnosis range that you might lose when you are

fixating on something that you see on a daily basis."

"It's always the top three or four differential diagnoses [that] spontaneously come to my mind. And we really start our treatments, management decisions based on those top differential diagnoses, and we only come to these rare diagnoses when we treat these patients and don't get a response. And that's when you start thinking about these things [CDSS]."

The biggest challenge in visually oriented medical specialty is the surprising similarity that different diseases may show to a less trained eye. Machine learning algorithms can easily spot such subtle differences in presentation, sometimes overtaking clinically trained specialists.

"Because in dermatology, unfortunately, we have the rules that are commonly presented, then there are so many exceptions. And the patients do not always present by the classic picture, as in the books."

To sum up in a participant's words:

"The hardest thing is to find the right diagnosis. Ten or five or 10% of patients that you think of at night, not knowing what, this [CDSS] can help them."

Reduce errors

"When I was a resident, I used to miss it all the time (in a low tone). It used to stand in front of me, clinically positive lichen planus, and I miss it. Then I started to pick my brain. Notice the small things are standing out as lichen planus in front."

Experts estimate that as many as 98,000 people die in the US alone any given year from medical errors that occur in hospitals (Kohn and Donaldson 2000). The statistics for most other countries are not available. CDSS could play a vital role in reducing this unnecessary loss of life. However, as discussed later, CDSS can potentially contribute to this problem too. The problem of misdiagnosis and its adverse effects were pointed out by an equal number of dermatologists and family physicians.

Most participants realize the importance of CDSS in reducing misdiagnosis. 62% of doctors who mentioned this had experience using a decision support tool in their practice.

"Broaden your mindset a little bit, so you don't miss those, you know outliers, because if I didn't think of common diagnoses as the centre of your bell curve, then the outliers which are less common may be missed."

A recent scoping review confirmed the effect of CDSS on reducing the rate of misdiagnosis (Muhiyaddin et al. 2020). The urge to reduce misdiagnosis is related to knowledge-seeking behaviour, which in turn is mediated by resource availability that is ensured by CDSS (Lai et al. 2014).

"And you don't want to make a misdiagnosis based on those common scenarios that you see every day. So, it stimulates you to look at those other things. Maybe at the end of it, it is still a common thing, and it may be an uncommon manifestation of a common thing, yes, but far more common than an uncommon thing."

In addition to patient safety, misdiagnosis can have economic consequences on the patient, as pointed out by one of the participants.

Really, it rubs off on a lot of cost and the patient's right to consider transportation coming to the hospital is going to spend more, and then medications that may be prescribed for the wrong reason.

Efficiency

I have summarized some of the common workflow related and training related concepts that emerged under the theme efficiency. The response related to CDSS impact on the health system efficiency was mixed. Some with previous experience of using CDSS complained about the increase in digital documentation and the negative impact on the patient-doctor relationship.

"Before when I had a piece of paper, I would be looking at the patient talking for 30 minutes and then the two minutes writing, or even looking and writing because we learned that at med school. Now, consultations are like maybe eight or nine minutes in the system and the encounters down to five minutes because they are not designed in a friendly way."

"I'm no longer a clinician, I'm a data entry clerk! I am asked to step away from my clinical practice."

When I am consulting with the patient, I am taking his history, right? With this kind of system [CDSS], I'm going to be reading most of the time, and you may lose focus.

Doctors' concern with the badly designed systems affecting their workflow and productivity is well known (de Grood et al. 2016). But with adequate training and adequately designed systems, CDSS could have a positive impact on efficiency, a sentiment echoed indirectly by few participants (Dikomitis et al. 2015).

"As they get to know the technology and they know how to use it very well. It should be like technology literacy first. It [is] for them to change their workflow."

"I think all doctors should be computer professionals and they have to have the knowledge of computers and also search engines and everything they do, go to different articles, different sites. So people are very comfortable navigating through

these systems."

System properties

"And you're seeing the patient, and you want to do things as quickly as possible, and then the system as you're trying to open a document it's like it'll take time to open. So if definitely speed is good so that will be easier, even want a perfect system, which is not possible."

A few participants mentioned some of the *context of use* related concepts such as the internet speed and system speed. Most respondents did not see it as a significant problem with the improvements in network speed and the technological advances in hardware.

"Nowadays, the internet is getting better. Especially in big hospitals."

Properties and dimensions

Some of the properties identified were active use (for decision making), learning use (for targeted learning) and leisure use (casual learning). To move towards theoretical integration (Urquhart et al. 2010), I sum up these themes under two easily recognizable dimensions along which these themes lie — clinic and hospital. The bidirectional arrow indicates the continuum of the level of specialization in a healthcare setting. A clinic is a small healthcare setting with one or a few doctors (mostly family practitioners) that caters to a wide range of low complexity diseases for all groups of patients. Clinics as the gatekeepers of the healthcare system are typically interested in reducing errors. A hospital represents a big setting, with many specialists, dealing with more complex diseases. Many hospitals are academic centres too, involved in the training of

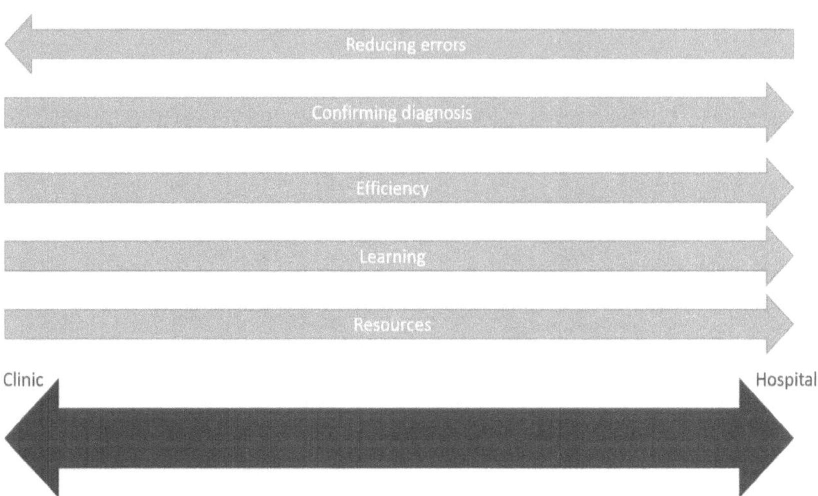

Figure 5.5: Dimensions of the 'use context' category

next-generation doctors. Hospitals typically have more access to resources and need to confirm diagnoses accurately in an efficient manner. The efficiency of hospitals that can be quantitatively measured using scales such as hospital readmission rate is a popular standard for assessing healthcare quality. The participant input discussed above and summarized in Figure 5.5 indicates that the knowledge needs of these two dimensions may be different, which is of great importance in studying CDSS adoption and design.

5.4.3   Information quality

Information quality is a widely used construct in IS research and can be broadly defined as the quality of the information content that is provided by an IS. I use this as a category to represent some of the themes that emerged, such as data source, as summarized in Table 5.4. Information quality has both theoretical and design aspects. For example, information quality provides theoretical insights into the knowledge needs of various users. At the same time, it provides prescriptive guidelines to designers, such as the credible data sources to use and privacy-preserving ML algorithms to use. Hence, I have included it as a combination category with contextual and prescriptive elements.

Table 5.4: Themes representing information quality

Category	Themes
Information quality	Data source
	Tailored information
	Need for evaluation
	Privacy

Data source

It was no surprise that data source emerged as the central theme, as in previous studies (Dinevski et al. 2011; Sim and Berlin 2003). The data source should be tailored to the type of knowledge requirement (Dinevski et al. 2011). DermML uses PubMed (Sayers 2010), a popular and reputed collection of peer-reviewed biomedical literature maintained by the National Center for Biotechnology Information (NCBI) as its primary data source. Many participants reiterated their trust in PubMed as well as the importance of using such credible and authentic data sources for machine learning. The method of extracting data from PubMed for the machine learning process in DermML is described in Appendix A.

"We have a sort of Trust. This [PubMed] is reliable information, evidence-based information, [that we] can all depend on."

"I'm glad that it's plugged to PubMed. That's the common site that people go to."

With the increasing popularity of ML and AI algorithms in CDSS, the credibility of data is a growing concern. In some specialties such as dermatology, the availability of credible data sources for image analytics is a big challenge, an area that I too have explored and made some suggestions (Eapen et al. 2020). Few participants echoed this concern:

"I think I would be more comfortable using solutions which are both based on evidence, and which are published, especially in machine learning solutions."

"It should be accurate. Yeah. I do believe in your need for more data. The more data you have, the more accuracy you have."

DermML adopts a crowd-sourcing strategy for image-based training and collaborative filtering for text-based ML . One participant raised a concern regarding her feeling about this strategy.

"If you're going to depend on your user, sharing the images to train the algorithms, I think it might not learn very well. Use publicly available [images], that you can teach your machine learning [on], not depending on the user."

Another prominent theme that emerged was the need to use multiple data sources. The importance of using additional data sources such as the systems biology data sources for corroboration has been proposed before (Yu 2015). Participants suggested Cochrane --- a collection of high-quality, independent evidence to inform healthcare decision-making — as an additional data source.

"Maybe other sites also like Cochrane, whatever evidence basis is present on

the internet relevant to that particular term."

Tailored information

Tailored information precisely conforms to the unique characteristics of a patient. Tailoring of messages is a facilitator for ehealth system use (Conway et al. 2017). This aligns with the IS value chain of data and information to knowledge that is vital for the success of any KMS (Alavi and Leidner 2001). Many participants reiterated the need for shortening the list of differentials. It is noteworthy that 74% of those who mentioned this need were family practitioners, the implication of which is discussed later.

"It gives me a very big list of things like whatever diagnosis. It should shorten my list to make sure that out of these 10 or 15, maybe it can give me five diagnoses, and I just want something like that."

"Differential diagnoses should narrow. It should not give me a large list."

Previous studies have highlighted the importance of "pre-analysis" of the large amount of data that CDSS systems can generate and potentially lead to information overload in the user (Ajami and Bagheri-Tadi 2013; de Grood et al. 2016). An interesting comment was the potential to do a self pre-analysis of the CDSS generated list to extract relevant information based on the clinical experience of the doctor.

"You start thinking as you already have the knowledge of dermatology, then you can say okay it cannot be an epidermoid cyst. You can put it all together because you have [done] clinical inspection palpation; you've done a lot of other things. So you can rule out. Out of those 15, at least 12 or 13, you narrow down to those two."

This aligns with my view (and a widely accepted view) that CDSS are knowledge augmentation systems that help the process of decision-making, rather than knowledge substitution systems that make decisions for the user (Dennis and Vessey 2005). Doctors are likely to accept systems that take their opinions and experience as input, an important design principle for DermML (Khairat et al. 2018).

Need for evaluation

The randomized control trial (RCT) is the gold standard for the clinical information of most clinical treatments. As CDSS belongs to the clinical domain, RCTs are commonly used to evaluate CDSS, comparing them against the doctors' decision as the gold standard (Sim and Berlin 2003). However, as I previously discussed, I doubt the relevance of RCT in testing knowledge augmentation systems.

"Just to prevent the bias, you should rather ask the clinicians first to list their differential diagnosis, and then show the system and see and then compare the results."

The need for clinical evaluation was mentioned by three participants out of the total 14. Though the number is relatively small, the views were rather forceful.

"It [DermML] is helpful, but how well the system works, it would only be determined when, as a clinician, they will test it in real practice, that's the real test. But as a concept, I think it's pretty good."

As many studies have shown before, doctors need proof of its utility prior to use (Chen and Hsiao 2012; de Grood et al. 2016; Dikomitis et al. 2015), a vital consideration for CDSS success.

"But I have not used it yet; it is difficult to say how well it works. So, I think it needs validation with the different physicians, going through it, goes into the iterations and going through them using this in practice, and then taking the input would be much more useful."

Privacy

Patient privacy and confidentiality is a growing concern among system designers and doctors alike. CDSS systems, primarily cloud-hosted ones, can be a vulnerability that can be utilized by hackers (Liu et al. 2018). As patient details that sometimes include pictures are used as input in most CDSS, there is a chance of inadvertently sharing sensitive information. Some participants echoed this concern.

"Inadvertently .... post a picture, and sometimes an identifying picture comes in, ... might be the lawsuit coming your way."

The responses from participants were not as strong as I expected, considering some of the high profile breaches that happened recently (Webster 2020). Most participants considered it a nuisance rather than an important aspect.

"In most scenarios, it should be okay, but unless it is a lesion on the face where the direct identification of the person .... That may be a tricky area, but most patients, if there's a consent form or anything they are pretty okay with the presenter, putting up their face, but some are not. And we have to respect that."

The need for approvals from regulatory bodies also emerged, and the difference in the privacy laws in different countries was highlighted.

"My only concern is that it is going to need FDA approval."

"Regulations, right? Different country different situations, right. That's

always a big challenge."

As mentioned previously, privacy is a theme that has a design aspect as well. The emerging privacy-preserving machine learning techniques is an area of active research (Mathew 2019). DermML adopts a privacy-preserving feature extraction method from images.

Properties and dimensions

In my view, most of the themes representing information quality seem to fall along the dimensions represented by information and knowledge. In simple terms, information is refined data, while knowledge is refined information (Alavi and Leidner 2001). Information tends to be more diverse and from multiple sources, less tailored to the patient, and may not be specific enough to be evaluated objectively or having privacy concerns. Knowledge tends to be actionable, tailored and specific enough to have privacy concerns. Because of the directly actionable nature, knowledge needs proper evaluation prior to use.

CDSS outputs may not be classifiable as one or the other, but it usually falls in a continuum between the two. For easy identification (and later theoretical integration), I symbolically represent the information quality of CDSS output as either information-rich or knowledge-rich, as shown in Figure 5.6.

5.4.4 Knowledge presentation

Knowledge presentation as a category represents the way knowledge is presented to the user. Common themes associated with this category are listed in Table 5.5. Some of the themes, such as user interface, are design-oriented. DermML adopts an uncommon knowledge presentation scheme where rare diseases are

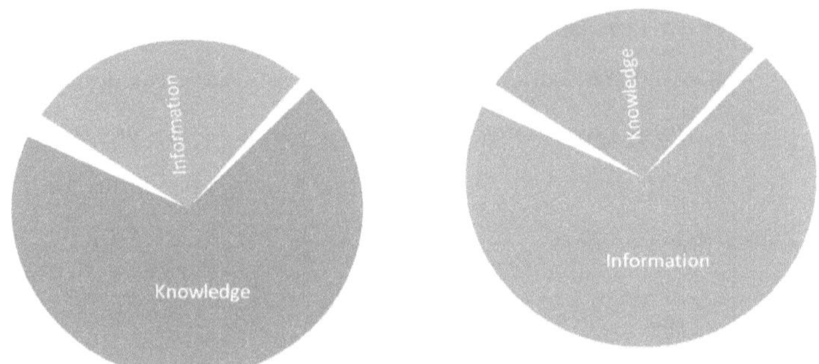

Figure 5.6: Symbolic representation of information-rich and knowledge-rich CDSS output

prioritized in the list of differential diagnoses. Knowledge presentation should be differentiated from the related, yet distinct concept knowledge representation — information capture in computer systems (Davis et al. 1993). As mentioned before, DermML uses RDF for knowledge representation.

Table 5.5: Themes associated with the knowledge presentation category

Category	Themes
Knowledge presentation	Prioritization
	User interface
	Collaborative filtering

Prioritization

Most diagnostic CDSS provide a list of recommended differentials. DermML adopts a unique prioritization strategy for this list. The rare diseases are listed before common conditions. This design logic is based on my personal observation that doctors use CDSS mostly to diagnose rare conditions.

Alert fatigue is a widely studied phenomenon in healthcare (Olakotan et al. 2020), and I assumed that a lot of accurate but irrelevant information might lead to reduced adoption. Some participants shared my view:

"Yes, These [CDSS] are usually used to detect rare conditions."

"You start with rarer cases; It's an important feature for this programme [DermML], and for the software [CDSS]."

But, an equal number did not:

"I think for me, I would rather have the most common diagnosis earlier in the list, rather than later. It would at least give the clinicians an idea or support their decisions that they are going in the right direction, or they were thinking about those common conditions before the rare conditions."

"You put the rare diseases first, because, like, there might be many rare [diseases] maybe one in 100,000, or one in like, you know million, or whatever."

The effect of prioritization has not been extensively explored except in the restricted context of drug-drug interactions (Cornu et al. 2018). Some doctors need the most likely diagnosis first, while others want to see the rare conditions, they are likely to miss. This shows the difference in the users' information needs, with the GPs favouring the first and the specialists preferring the latter.

User interface

The user interface is a well-known factor in the adoption of any IS, and CDSS is no exception (Miller 1990). There are many usability studies on the various aspects of user interfaces and user experience (Carroll et al. 2002; Ying-Jui et al. 2003; Yuan et al. 2013). As expected, many opinions surfaced during the study from the participants, of which a few interesting ones are listed below:

"I don't understand what the difference between those two columns is."

"The interface is not good. But it is going to improve in the final version, right?"

"As a testing platform, it looks great. So, you don't have to worry about it."

"The user interface needs a bit of cleanup. (laughs)"

"Generally, the buttons on the test button should not be that big. You put two columns next to each other. Yeah, that's a bit confusing for a lot of people. You can make it a single page."

A couple of participants mentioned the importance of colour coding, a useful technique for semantically grouping similar information.

"Right now, I'm looking at multiple diagnoses. But if, we could have some kind of colour coding based on maybe importance or rarity, it may make it easier."

"Maybe like colour coding will help me like when I'm opening and seeing you know maybe the 10 or 20 or 30 lists of diagnosis."

As previously described, *knowledge presentation* as a category has both contextual as well as design-related themes, and the user interface is a theme with specific prescriptive design information.

Collaborative filtering

Collaborative filtering as a method has a specific meaning in recommender system design. However, I have used it as a theme here in a broad sense to represent ranking and filtering of clinical rules, learnt using machine learning methods, collaboratively by users (Terveen and Hill 2001). As detailed in Appendix A, DermML uses collaborative voting to calculate the weight of rules in decision making.

Personally, I used to consider this the highlight of DermML, making it one of the few ones that learn from the cumulative experience of doctors. To my surprise, the response to this feature was lukewarm at best:

"The upvote and downvote. that's a good thing."

And dismissive in general:

"I would prefer it to be more related, as you said to mathematical or statistical studies on actual patients versus more of a subjective opinion of the doctor saying, okay, I, I haven't seen it an oral mucosa. So I'm not, I'm downvoting it, or I'm deleting it."

This has implications on design, as discussed in the design theory later.

Dimensions

I adopt the constructs, *tacit knowledge* and *explicit knowledge,* from knowledge management (Alavi and Leidner 2001) as the two dimensions of knowledge presentation. Tacit knowledge in KM is the knowledge embedded in practice that is difficult to externalize and transfer, and the explicit knowledge is the opposite (Polanyi 2009). I have used it as a dimension of *knowledge presentation* in a related sense, but with some difference. I have used tacit knowledge to represent knowledge that is commonly known to everyone in a specialty. In a list of differential diagnoses, these are the first diseases that come to mind. Explicit knowledge digitized in journal articles may not be apparent to everyone.

5.4.5 Technology efficacy

I have used technology efficacy as a category to represent some of the design and technology-centric themes that emerged. It is crucial to differentiate technology self-efficacy (sometimes abbreviated to technology efficacy in literature) from this category. Technology self-efficacy — "the belief in one's ability to successfully perform a technologically sophisticated new task" (McDonald and Siegall 1992) — is a property of the user. I have used technology efficacy to represent system properties. The themes under this category are listed in Table 5.6

Table 5.6: Themes grouped under technology efficacy

Category	Themes
Technology efficacy	Participatory design
	Integration
	Packaging
	Algorithm
	Continual learning

Participatory design

The importance of requirement elicitation in systems design is widely recognized. In the agile methodology, user input is sought at every stage of iterative design (Humayoun et al. 2011). Participatory design goes a step further and seeks the involvement of users in the design process itself (Donetto et al. 2015). As technology is becoming pervasive in healthcare and eHealth hybrids (Heeks 2006) — doctors turned techies — become common, more and more doctors would be interested in getting involved in participatory design. Some of my participants did mention this emerging trend or complained about the lack thereof:

"I feel there's a new generation of people like myself who are willing to take

part."

"Physicians are not at all involved."

Integration

Integration with other HIS is fundamental for CDSS success (Kux et al. 2017). This emerged as a prominent theme with three participants stressing the need for integration with electronic health records (EHR), and a couple suggested integration with drug formularies.

"For this, I think the first thing in my mind was integration to the EHR because that's very important because if it is separated from EHR, it's pretty disruptive."

"Many things could happen, like a doctor may prescribe something that already patient is on it, like a medicine and he may duplicate it, or made a mess of the dose, or increased the dose unintentionally, this kind of stuff like medication error can happen. And I think CDSS can help them. Flag things like this as duplicate or the dose is higher than the normal range and this kind of stuff, so it can help make sure patient safety."

Packaging

A couple of participants with experience practicing in low resource areas mentioned the need for a desktop version and a mobile version. Because of the unpredictable internet connectivity in some regions, a web-based CDSS may be unstable. Doctors may not have access to a computer in some situations, and a mobile version will come handy.

"A wired [networked] desktop version is [could be] more stable, and the better and therefore, the system faults are much less."

Algorithm

I was not expecting any insights related to the machine-learning algorithms used by DermML as the participants were not experts in IT. One of the comments from a participant solved a long time design dilemma that I faced. Image classification in general dermatology is highly inefficient because of the large number of output classes, 4144 in the case of DermML. One of the participants suggested an initial doctor-led classification into a broad disease category so that further machine classification becomes easier and more accurate. This response proves that qualitative research can contribute to actionable design knowledge.

"There could be some conditions that we can start with not all dermatology but certain conditions, for example, papulosquamous or bullous, just the bullous disorders and train on bullous, so categorize, for categories like we used to teach our residents, bullous diseases, how we classify them and then classifying [or] trained by pictures and images, bullous alone, maybe papulosquamous alone, maybe pigmented lesions alone. That is not something that is across the board because it's impossible to hone in on those differential diagnoses of certain conditions that look similar."

Continual learning

As I mentioned in the introduction, medical knowledge keeps changing as new evidence emerges from clinical research. It is vital for CDSS to keep up-to-date as it has grave patient safety implications (Nwengya et al. 2011). The need for continual learning systems was emphasized by a couple of participants:

"I think a part of it, which is based on realtime data from PubMed or the other major databases. I think it [DermML] could not have problems, but in general,

some of these CDSS have problems because they need to be maintained regularly and up to date."

"[I would use DermML] rather than on a system that's trained, like, two years back or three years back, especially with pandemic conditions and newer conditions coming, that the world is facing, it's better to have a more realistic, machine learning-based CDSS."

Both participants recognized the continual learning approach of DermML as a strong point in favour of their intention to use.

5.5   Clickstream data

As mentioned previously, I recorded anonymous clickstream data from the site. As the clickstream data used a custom script that filters all individually identifiable data, including IP address, a detailed user-level analysis was not possible. However, the clickstream data did show some useful findings. A total of 780 interactions with the system was recorded during the study period, most of which were page views (520). Participants clicked on the links 195 times. Testing the diagnoses function (see artifact design in Appendix A) with a simulated patient history initiates a 'change' event and searching for diseases initiates a 'submit' event. Change and submit events were 54 and 11, respectively. The clickstream statistics are summarized in Figure 5.8

DermML has information on 4414 distinct dermatological conditions. The participants only explored 28 distinct diseases. The help page was only accessed twice during the study period.

The clickstream data shows that the exploration and engagement of the

1st Order Concepts	2nd Order themes	Category
Searching can immediately give idea on diagnosis. Recollect forgotten knowledge. Keep moving on after diagnosis	Cognitive load	
Continuing medical education Leisure learning Patients challenge knowledge	Learning new information	
They depend more on experience. Older doctors do not use. Easy access to information	Practitioner	System User
People are resistant to change. They like to do what they are used to doing. Doctors are reluctant to use new technology	Change Management	
Automation is coming to replace me. Patients require human contact. Like to be in control	Threat to professional autonomy	
An internal social media platform for healthcare workers Patient information inadvertently shared More open to criticism	Socio-cultural	
Low-resource vs High-resource Academic vs non-academic Hospital vs clinic diagnostic vs therapeutic	Healthcare setting	
Second guess diagnoses Fixating on common diagnoses Presentation is not always classic	Confirm decision	Use context
Avoid missing outliers. Easy to forget rare things. Save resources	Reduce errors	
Change of workflow. Doctor-patient engagement Becoming data entry clerk	Efficiency	
Internet speed System speed	System Properties	
Reliable, evidence-based information Should be accurate. Use multiple data sources	Data source	
Narrow list of differential diagnoses Should be based on patient characteristics. Self-filtering based on clinical experience	Tailored information	Information quality
Prevent bias. Test it in real practice. Validate with different doctors	Need for evaluation	
Inadvertently posting pictures Regulatory approvals Need for patient consent	Privacy	
Rare diseases first Common diseases first Length of list	Prioritization	
Semantic coloring Combine related information together. Buttons and columns	User interface	Knowledge presentation
Upvote and downvote Learning from user contributed images	Collaborative filtering	
Involve doctors	Participatory design	
Integration with EHR Integration with drug formulary	Integration	
Desktop version could be more stable	Packaging	Technology efficacy
Do not start with all conditions, but categorize	Algorithm	
Based on real-time data More realistic especially for pandemics	Continual learning	

Figure 5.7: Analysis flowchart

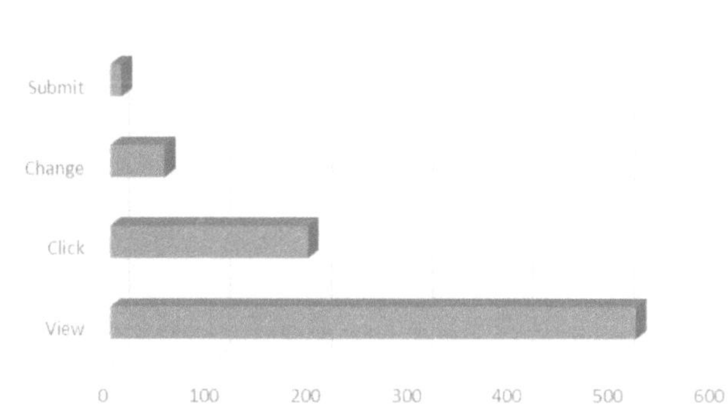

Figure 5.8: User interaction with DermML

participants with the CDSS were limited (Brodie et al. 2015). The participants attributed only limited value to the information provided by DermML.

5.6 Theoretical coding

Having identified the core categories and backed by my reflexive memos, I now proceed to infer the relationships between these constructs --- the process of theoretical coding. As mentioned previously, I have used the first four categories — system user, use context, information quality and knowledge presentation — at this stage. The last category — technology efficacy — is for design knowledge and would be used separately for proposing prescriptive type 5 theory.

I use my data, personal experience and available literature during this process. Data includes the demographics of the participants and the transcripts, which includes the participants' perceptions about the other groups. The system user's dimensions and the use context create four quadrants for CDSS use with different knowledge needs as depicted in Figure 5.9.

The overarching theme of this substantive theory is that *doctors are not a homogenous group, and different doctor groups have different knowledge needs.* Though this may seem obvious in hindsight, this is rarely recognized in practice. The practical significance of this proposition in the design and implementation of CDSS is enormous. Below, I describe the relationships in detail and their practical significance.

A specialist in a hospital needs immediate access to information: A specialist is trained to contextualize information to the patient, and they rarely trust contextualization by a CDSS. Professional autonomy is valued more by this group and will resist any perceived threat to autonomy. Hence, in such scenarios, a CDSS should provide as much relevant information as possible, without trying to filter it or contextualize it. Evidence is of high importance. Providing anchors with colour coding and semantically grouping related information would be appreciated by this group.

The threat to professional autonomy and the importance of learning new information was mentioned more by those with specialty training than by family physicians. I have personally experienced this need while practicing as a specialist in a hospital. For this group, the knowledge is embedded in the user with the CDSS providing condition for easy access to information (Alavi and Leidner 2001; McQueen 1998).

A GP in a clinic needs contextualized knowledge: GPs serve as the vital gatekeepers of the healthcare system in most countries, including Canada (Liang

Figure 5.9: Schematic diagram of CDSS contexts and purposes

et al. 2019) and as such, require efficient access to tailored, timely, patient-specific knowledge (processed information) that applies to the majority of cases. They are less interested in exceptions to the common rules and rare conditions. They practice in conditions where system properties and the user interface are of greater importance.

As mentioned before, saving time emerged as an important theme, supported by more family physicians than dermatologists.

"Time is very precious to both the patient and to the clinician."

Some participants who were specialists discussed some of the potential problems that GPs can face:

"In dermatology, there is a lot of (laughs) a large number of diseases that have the same features."

The differences in priorities between primary and secondary care physicians have been reported previously in a qualitative study based on focus groups (Varonen et al. 2008). In this context, where the number of tasks, duration and number of instances where system use is required is high, the perceived ease of use is correlated with system use (Igbaria and Zinatelli 1997).

Residents seek to reduce errors: Medical residents are doctors in training. Many participants highlighted the knowledge needs of residents, and as such, CDSS (DermML) was seen as an ideal tool for them to learn.

"Residents during the case presentation in the ward, they put [patient details] on this [CDSS], [and they can know] what are their differentials when the consultant asks."

Residents aspire to reduce errors during the training period and seek to internalize knowledge commonly known to senior colleagues. Hence, residents need easy access to standard information, and their dependence on the system use is related to the information quality (Ravichandran and Rai 1999).

Specialists in a clinic seek explicit knowledge: Specialists have the tacit knowledge required to manage common cases. Specialists in a less challenging situation may indulge in leisure learning. During such episodes, specialists seek uncommon knowledge contextualized to patients for which a definitive diagnosis is in doubt. Many specialists among the participants echoed this sentiment.

"10 or five or 10% of patients that you think of at night, not knowing what, this [CDSS] can help them."

The importance of these relationships in predicting the CDSS adoption is elaborated in the discussion section. Next, I propose a type 5 prescriptive theory for the design of CDSS.

5.7   Design theory for CDSS

To guide CDSS design, I propose a set of prescriptive guidelines based on the themes related to technology efficacy. The guidelines are summarized in Figure 5.10.

Involve doctors early in design. As previously described, participatory design emerged as a significant theme. Doctors tend to trust the system more when they are involved in the design. The lack of inclusiveness among CDSS was a major complaint.

Link outputs from humans and machines. Machine learning algorithms should cooperate with humans whenever possible. One example that surfaced during the study was the potential for humans to do an initial high-level classification, before the machine-learning-based final classification. This reduces the number of output classes and improves efficiency.

Follow standards for better integration with other HIS. Standalone CDSS should integrate with the HIS that doctors already use. Doctors hate to login to multiple systems and to deal with many screens to get the required information. Emerging integration standards, such as *SMART on FHIR* (a standards-based, interoperable apps platform (Mandel et al. 2016)), should be explored.

Utilize edge computing whenever possible. Lack of good internet connectivity is still a problem in low resource areas. Doctors prefer desktop and mobile applications. Desktop applications give doctors a sense of control over their data.

Plan for continual-learning and easy maintenance. CDSS are evolving systems that should be continuously updated as new evidence emerges. This is vital for patient safety. CDSS should adopt a continual-learning strategy or an implementation that allows easy maintenance.

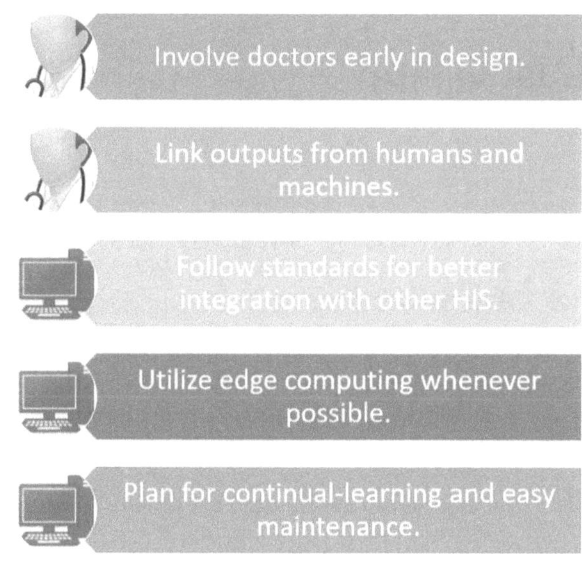

Figure 5.10: Prescriptive rules for CDSS

Chapter 6

Discussion

"nothing is so practical as a good theory." - Lewin (1945)

6.1   Introduction

The objective of my research was to find the factors responsible for user adoption of diagnostic CDSS in clinical practice, a problem that is currently less explored. I used a real stimulus that I designed to elicit a response from my participants. My secondary objective was to gather prescriptive knowledge to improve CDSS design.

What I found was that the information needs of doctors differ according to their individual characteristics as well as the context of the use of CDSS. The systematic analysis of what my participants told me, and my own experience showed that the information needs could be explained and predicted within certain boundaries.

The rest of this section is structured as follows. First, I discuss the significance of my findings with respect to the research questions, I set out to investigate. Next,

I explain some of the categories that emerged during coding, in the context of
existing theories in IS. I represent the relationship between the categories in Figure
6.1. Finally, I discuss the level of conceptualization of the proposed theories,
limitations, and recommendations for future research.

6.2 Significant findings

I started off by asking the overarching question, ***how do doctors describe and
characterize their use of CDSS in their practices and their views on
how CDSS should be designed***? The overarching finding of the study is
that doctors characterize their use of CDSS as an augmented learning tool or
a decision support tool depending on the context and type of practice. CDSS
designers should consider the knowledge needs of the specific group they intend
to support and involve them early in the design process to improve adoption.

What are the characteristics of individual CDSS users?

At a high level, most existing CDSS adoption studies focus on internal factors such
as personal expectations and perceptions of users and their moderators such as age
and gender, separately from organizational and technical factors (Mah et al. 2020).
I found that external factors such as the practice setting, and type of practice
influence these internal factors. Most CDSS adoption studies derive from theories
and models of adoption from IS such as UTAUT (Venkatesh et al. 2003) and
the DeLone and McLean model (Delone and McLean 2003). UTAUT constructs
such as PU and PEOU were identified as key factors facilitating CDSS adoption
(Sambasivan et al. 2012). My focus on the characteristics of CDSS users led me

to the insight that GPs, residents, and senior doctors have different expectations and perceptions about the effort expectancy and ease of use. This relationship has been previously suggested in the context of primary care (Lugtenberg et al. 2015; Short et al. 2003). Experienced doctors are believed to be resistant to CDSS adoption (de Grood et al. 2016; Leslie et al. 2006). However, this did not emerge in my study and I posit that such differences in adoption may be related to a misfit between the type of CDSS available and the knowledge needs of experienced doctors.

How do doctors use CDSS in practice?

I explored how doctors use CDSS in their practice. I identified three distinct types of use: (1) active use (for decision making), (2) learning use (for targeted learning) and (3) leisure use (casual learning). I found that the users' learning style depends on external factors such as the practice setting. Active use for decision making happens in hospitals where the patient load is high. Residents commonly involve in targeted learning while specialist doctors in less challenging circumstances (as in a clinic) engage more in casual learning. To my knowledge, this distinction has not been described before.

Does the scarcity of resources impact CDSS use?

The impact of the availability of resources, an initial focus of my inquiry did not yield any insights. The cancellation of the planned secondary site for data collection because of the COVID-19 pandemic may be a factor as I was expecting insights on this aspect from there. The CDSS adoption in resource-poor areas may vary because of the difference in the information needs of doctors practicing in such settings. Previous studies have suggested this relationship (Raghu et al. 2015) but this was not a major concern in my participants with resource-poor practice experience. The effect of resource availability on CDSS adoption still needs further exploration (Ahlan and Ahmad 2014).

What are the factors associated with enhanced CDSS use?

CDSS adoption and its use are known to be correlated with computer literacy (Wilson and Opolski 2009), support and training (Lai et al. 2006). However, this did not emerge as a prominent theme in my study. Previous studies have suggested the differences in CDSS adoption among young doctors (Tsiknakis and Kouroubali 2009; Chan et al. 2004). My findings suggest that such differences may also be related to the variations in the knowledge needs of residents and young clinicians. Congruence of knowledge needs and the ability of the CDSS to satisfy the specific need emerged as the most important factor responsible for enhanced CDSS use and its subsequent adoption.

What are the design factors that affect CDSS use?

A frequently reported barrier to CDSS adoption is the lack of confidence in the CDSS knowledgebase (Khairat et al. 2018; Shibl et al. 2013). This aligns with the need to involve doctors early in the CDSS design, a vital theme that emerged from my study. The importance of the knowledge presentation format in system design from my study is in agreement with previous studies (Belard et al. 2017; Moxey et al. 2010).

Next, I discuss the categories that emerged during coding, relating it to the existing knowledge in the IS domain.

6.3 Categories and their significance

The five categories that emerged from this study were: (1) system user (2) use context, (3) information quality (4) knowledge presentation and (5) technology efficacy.

6.3.1 System user

Szajna and Scamell (1993) described the effect of system user and their expectations on the success of IS implementations. The study used the cognitive dissonance theory to explain the behaviour of system users with certain expectations of the system. This aligns with my findings that assessing user expectations and needs is vital for CDSS adoption. One of the commonly discussed factors under this category in the CDSS literature is the threat to professional autonomy (Esmaeilzadeh et al. 2015; Sambasivan et al. 2012; Walter and Lopez 2008), a factor that I found important as well. Characterizing the various dimensions of this category such as GPs, residents and specialists and relating them to their knowledge needs is one of my important contributions.

6.3.2 Use context

Context is an important factor in IS design that traditional requirement analysis methods fail to capture (Cherry and Macredie 1999). Use context is generally studied in the organizational perspective in IS (Avgerou 2001). Use context in CDSS has prominent personal factors such as the cognitive load of the practitioner and the type of practice, in addition to organizational factors. Understanding these factors is important not only for designing successful systems but also for enhancing patient safety (Faiola et al. 2015).

6.3.3   Information quality

Information quality is an important construct in the DeLone and McLean model indicating the desirable properties of the output (Delone and McLean 2003). In the context of CDSS, PU and PEOU mediate the effect of information quality on adoption (Tao 2008). The importance of data sources and the need for tailored information that emerged in my study have been demonstrated previously (Dinevski et al. 2011; Sim and Berlin 2003). Privacy is an important concern in CDSS (Liu et al. 2018). However, my findings did not reflect any significant privacy concerns despite some recent breaches (Webster 2020) that caught public attention.

6.3.4   Knowledge presentation

This category includes the form and method of disseminating information to the end-user (Kelton et al. 2010). Though this concept has been extensively explored in IS, CDSS studies have given this concept less importance compared to the related concept "knowledge representation" (Davis et al. 1993). I investigated the effect of factors such as the length of the list of differentials presented to doctors and the order in which the differential diagnoses are presented. As mentioned previously, I found that they also depend on the practice type and setting. This is in agreement with the IS literature that shows the relationship of presentation with user involvement (Lurie and Mason 2007) and with task knowledge (Mauldin and Ruchala 1999).

6.3.5 Technology efficacy

As previously mentioned, I have included disparate findings with design
implications under this category. The participatory design (Donetto et al. 2015)
and the integration with other HIS (Kux et al. 2017) are widely accepted as
success factors in CDSS design. I also found that the packaging (cloud-based vs
local installation) emerged as an important design factor in addition to
participatory design and integration. Surprisingly, local installation was
preferred over cloud based deployment. This may be related to a "mismatch
between doctors' conceptual understanding of cloud computing for health care
and the actual phenomenon in practice (Gao et al. 2018)."

Doctors can potentially utilize up to two million pieces of information to
manage patients (Smith 1996). It is difficult to develop CDSS systems that
satisfy all the information needs of doctors. Most CDSS systems have utility
only among certain user types and practice settings. Introducing CDSS in a
setting different from what it was primarily designed for can lead to suboptimal
adoption. Traditionally, system and information quality are assessed along with
the PEOU and PU in investigating the CDSS adoption (Chen and Hsiao 2012).
My findings suggest that the user may seek tacit knowledge, explicit knowledge,
raw information or contextualized knowledge depending on the user type (GP vs
specialist) and use context (hospital vs clinic). A CDSS may satisfy some of
these needs depending on how it presents the knowledge to the user and the
technology utilized. The adoption of CDSS depends on the congruence of user
factors with the system factors. This relationship is summarized in Figure 6.1.
To my knowledge, the variation in the information needs of doctors depending
on their type of practice has not been previously investigated in the context of

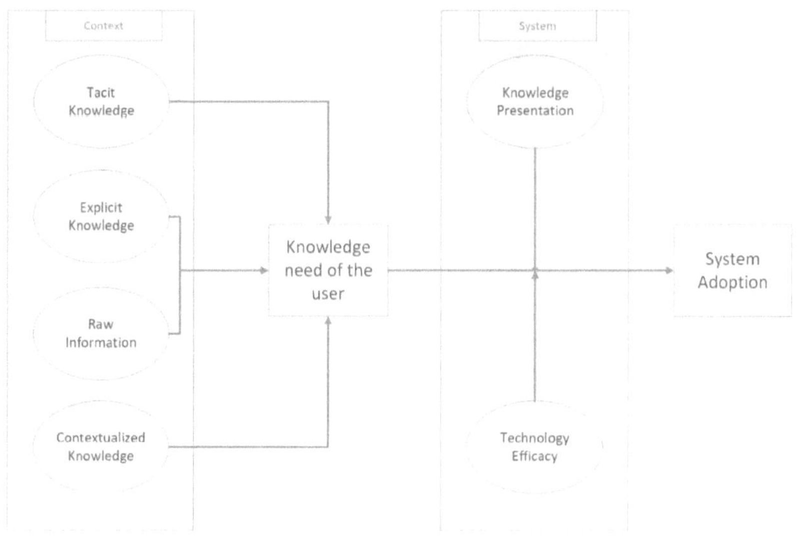

Figure 6.1: Model describing CDSS adoption by doctors

CDSS.

6.4 Theoretical integration

I adopted the (Charmaz 2006) constructivist school of GT (co-creating reality in a group to which I belong) with the (Gioia et al. 2013) style of methodical analysis (from 1st order concepts to 2nd order themes to aggregate categories and their dimensions) and scaled up the theory according to (Urquhart et al. 2010) recommendations for theorizing in GT. I have scaled the theory beyond the bounded scope of dermatology to the substantive scope of all medical specialties. The findings from the interview transcript were scaled from descriptions to a theory that identifies core categories and their implicit relationships. This

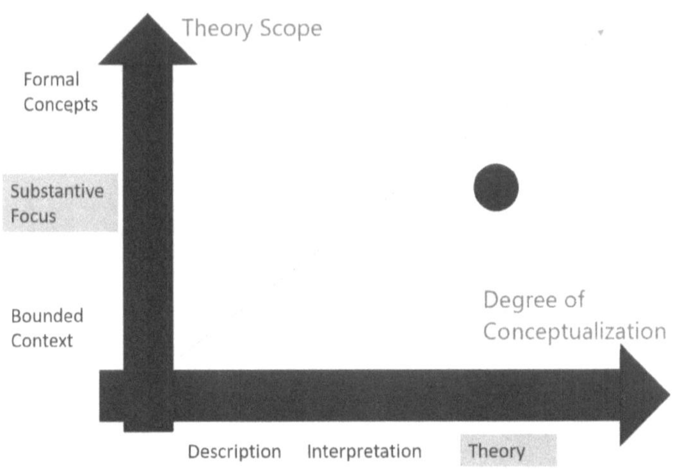

Figure 6.2: The scope of the theory of CDSS adoption

substantive theory of user adoption of CDSS belongs to the type 4 IS theory for explanation and for prediction in Gregor (2006)'s typology. The theory can explain suboptimal adoption of CDSS (when it caters to a particular knowledge needs but is used in a different user segment) and predict the outcome based on the congruence of user segment and knowledge needs. Figure 6.2 depicts that the scope of the explanatory and predictive theory for CDSS adoption that I propose has a substantive focus on CDSS according to Urquhart et al. (2010) typology.

The final process in GT theorizing is the theoretical integration — relating to other kernel theories in the specialty. The theoretical coding has scaled the theory beyond the limited scope of dermatology, as it postulates relationships between the specialty-independent categories as described before. Then by using the notion of 'knowledge-seeking' behaviour of the doctors under different contexts, the theory integrates with the existing theories of knowledge management and knowledge-seeking behaviour of professionals. However, my intention is not to scale theory to the level of formal concepts in IS, as the gap and the need for a theory is at the substantive level of CDSS.

Alavi and Leidner (2001) proposed the popular model of knowledge management in the context of firms. The model describes the relationships between the tacit and explicit knowledge of individuals, and the role IS play in their interconversion. The model also describes the importance of the internalization of the episodic memory of groups into semantic memory, an essential consideration in CDSS. My findings show that the importance of episodic and semantic memory differs according to the context.

The model I propose also aligns with the up-front theory (TPB) (Ajken 1991), especially in the importance of the constructs — attitude and PBC. However, testing the up-front theory was not my intention. A previous study that used TPB to study the knowledge-seeking behaviour of professional virtual communities concluded that "knowledge-seeking behaviour is solely determined by knowledge-seeking intention" (Lai et al. 2014). The same study showed that system quality and resource availability are the determinants for PBC of knowledge seeking. The Montazemi et al. (2012)'s study on factors impeding knowledge transfer discusses the importance of the 'realized absorptive capacity' of the recipient for new knowledge, an important consideration in the system use dimension. Another study showed that knowledge sharing could only take place when both knowledge contribution and knowledge-seeking exist (Bock et al. 2006).

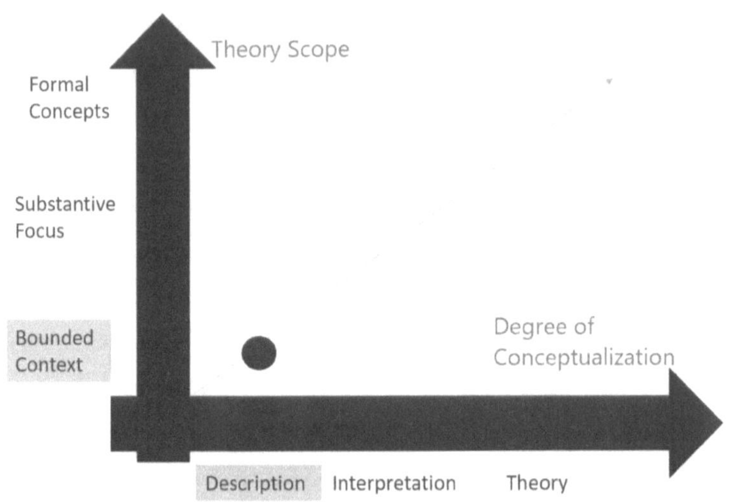

Figure 6.3: The scope of the proposed CDSS design theory

6.4.1 Design theory

Design theory belongs to the type 5 theory for design and action according to Gregor (2006)'s typology and, as such, does not attempt to scale beyond prescriptive guidelines (Figure 6.3). In the Urquhart et al. (2010) typology, the design theory is a set of descriptive statements within the bounded context of CDSS.

6.5 Limitations

The dimensions of system user — GP and Specialist — in this study did not imply chronological or professional seniority. The inclusion of residents with GPs was to help in the scaling of theory and does not mean that they have identical roles.

Similarly, reducing the healthcare context to the hospital <-> clinic dimension is to ensure parsimony. The healthcare setting is inherently complex and multi-dimensional, so that there would be some loss of representational accuracy during theoretical coding.

DermML as a stimulus is primarily designed for diagnostic decision support in dermatology — a specialty that gives importance to visual cues. It may not have elicited some responses vital in understanding the decision-making process under challenging settings such as an ER. This is an area of active research to which this study does not substantially contribute.

As I conducted all the interviews, my theoretical sampling may have been influenced by my contacts and access to resources. Though I have taken care to minimize the effect by strictly following a theoretical sampling process, and I reached theoretical saturation, I cannot guarantee that all views on this topic are adequately represented. Besides, the participants identifying me as a professional colleague may have blunted some of their criticisms.

There was a minor change in protocol due to the COVID-19 pandemic leading to initial interviews being face to face and the rest virtual using Zoom. Personally, I did not notice any difference or encounter substantial challenges due to this change.

Clickstream data is highly useful for behaviour analysis when combined with other forms of data. However, I did not collect individual-level data as recommended by the ethics board and, as such limited a detailed analysis linking various types of data. Image analytics was not available for participants because of privacy concerns, and the participants based their responses on the hypothetical future state, as explained during the interview.

Dermatology, as a medical specialty, has diverged significantly of late with the emergence of cosmetic dermatology as a subspecialty with distinct characteristics beyond the scope of my inquiry. Though the sample has a small representation from cosmetic dermatology, I have not made any detailed analysis of the differences or its impact.

6.6   Problems faced

Other than the minor protocol change necessitated by the COVID-19 pandemic, I conducted most of the interviews during a tough time for healthcare workers. A multi-site comparison of user adoption was the original plan, but pandemic related issues necessitated the cancellation of the planned site in India.

6.7   Implications of the findings

My study offers suggestive evidence for heterogeneity in CDSS users and context. Most CDSS evaluations, adopting an RCT methodology or systematic reviews, are not designed to incorporate this heterogeneity. This may be one of the reasons for the many failed CDSS implementations and the widening rift between doctors and IT experts who fail to involve them while designing such systems.

GT is emerging as a popular qualitative method that is capable of handling contrasting worldviews. This makes GT an ideal method of behavioural inquiry related to IS artifacts. Though there are previous attempts at combining GT with DSR, a well-defined method of combining both is not available to my knowledge. This study offers a pragmatic method to use GT for theorizing and design. I call this the real stimulus method (RSM). To guide future researchers interested in adopting this method, I offer some prescriptive recommendations:

6.7.1 Recommendations for future research

I have the following recommendations for researchers adopting the real stimulus method (RSM) in GT exploration of an IT artifact:

1. Scale the artifact from the level of a construct, model or a method to a tangible implementation and use it as the stimulus.

2. Base the design of the stimulus on a design theory and use the same theory as the sensitizing theory in the GT method. Typically, the theory should guide the interview guide and analysis.

3. Demonstrate the stimulus prior to the interview and, if possible, grant the participants access to the artifact for exploration. If participants have access, record the clickstream data.

4. During the demonstration, make the aim of the study clearly known to the participants, emphasizing that evaluation of the artifact is not the primary objective.

5. Try to separate contextual concepts and design concepts during coding. Some concepts may be a combination of both. Maintain this division in the 2nd order themes and in the aggregate categories.

6. Look to propose different theories to explain/predict behaviour and guide design. The design theory may be a set of propositions, and the contributions may be at a construct, model or method level and not necessarily at the instantiation level (Baskerville et al. 2018).

Though the RSM that I propose has similarities with user-centred design and prototyping — the process of implementing ideas in the tangible form for testing — RSM has important differences from both these methods. First, RSM systematically collects user feedback using GT, which has its own set of guidelines and rules. The genre of user-centred design methods focuses mostly on the user interface design, but RSM collects a much more comprehensive range of information, including refinement of the algorithm — the steps followed in reaching the output. Besides, RSM aims to build generalizable design knowledge that is relevant beyond the IT artifact. RSM is likely to be more useful during the initial stages of the design process.

Prescription alerts and drug interaction checkers are more widely used CDSS than diagnostic CDSS such as DermML. Future research could focus on these CDSS and assess their user adoption. Another avenue for future study is the effect of patient-reported outcomes as CDSS input and its effect on the decision process. It is also important to investigate patient-related factors in the user adoption of HIS in general and CDSS in particular.

6.7.2   Recommendations for future policy

The World Health Organization has identified "sound and reliable information is the foundation of decision-making across all health system building blocks, and is essential for health system policy" (WHO 2014) and CDSS plays an important part in its realization. This study indicates that the information needs of doctors and the context within which doctors practice medicine vary considerably, and no single system can satisfy the needs of all. The study proposes that policymakers take the peculiarities of the information need of various regions and choose a tailored HIS strategy that considers context.

The accuracy dimension of information quality is the basis for most CDSS evaluations conducted by clinical research teams. A comprehensive evaluation framework is needed to assess the system success of CDSS. Generic IS success measurement instruments may not be suitable for this purpose (Sedera et al. 2004). I propose some recommendations for evaluating CDSS based on this study:

- System quality measures should include maintainability, integration and user-friendliness.

- Information quality measures should include relevance, currency and format. The format depends on the user segment in which CDSS is used.

- Always include an evaluation of net benefit from a doctor and patient perspective. Assess the access to new information, increase in productivity, and decision effectiveness in addition to decision accuracy. Measure any improvement in patient outcomes at the group level and individual level.

6.7.3   Recommendations for future action

The study makes recommendations for CDSS designers with the aim of bridging the widening gap between doctors and designers. The overarching message to the design community is to involve doctors in every phase of design, from algorithms to the user interface.

6.8 My contributions and reflections

This study was an incredibly insightful and humbling experience for me. Despite receiving multidisciplinary training in health and IT, I never recognized the difference in the various healthcare teams' information needs. I was aware of the importance of participatory design, but I never realized that users could give insights beyond the user interface design. I was fascinated by the ideas for improving the machine-learning algorithms that I used for building DermML .

The artifact design completes my research in this domain that I started more than a decade back. I began by proposing a domain ontology for dermatology (Eapen 2008) and followed it up with a field-tested method in Columbia (Sáenz et al. 2018). DermML is the final instantiation extending my previous work to the exciting and emerging world of multi-modal machine learning and artificial intelligence in healthcare.

Chapter 7

Conclusion

CDSS have a history extending for more than five decades. Despite some successes in specific domains, broad adoption and widespread use of CDSS in clinical practice have not been achieved (Greenes et al. 2018). The existing theories of technology adoption have been unsuccessful in describing HIS adoption in general and CDSS adoption in particular (Hu et al. 1999). There seems to be a paradoxical non-correlation between intention to use, and actual system use in CDSS.

I found that the information needs of doctors vary, based on their type of practice and the healthcare setting in which they practice. Most CDSS do not provide all information required by doctors. The suboptimal adoption of CDSS by the clinical community may be due to the disparity between information needs and information availability. The learning style of the CDSS users and the type of use (leisure learning, active use etc.) depends on their information needs. The scarcity of healthcare resources affects adoption as the information needs of the doctors depend on resource availability. Involving doctors early in CDSS design is critical.

We are going through an era of unparalleled technological advances in health informatics. Machine learning and artificial intelligence-based healthcare applications are gradually becoming ubiquitous. These technologies have the potential to improve healthcare delivery and reduce healthcare expenditure.

The new information technologies and knowledge management systems have widened the chasm between IT professionals and clinicians, both grappling to understand, support and utilize each other. Heeks (2006) emphasizes the importance of clinicians who are e-health hybrids making a transition from healthcare to IT and have the capacity to understand both worlds. The IS field has contributed significantly to the development and application of information technologies in healthcare's economic, social, and organizational aspects (Chiasson and Davidson 2004). I hope that my experience as an e-health hybrid and the rich insight I gained from this study will help me to design successful CDSS and guide other CDSS designers.

www.ingramcontent.com/pod-product-compliance
Lightning Source LLC
LaVergne TN
LVHW042158190726
843493LV00006B/1722